APR 2 5 2022

PORT WASHINGTON PUBLIC LIBRARY
ONE LIBRARY DRIVE
PORT WASHINGTON, N.Y. 11050
TEL: 883-4400

ALBERT EINSTEIN
THE MAN, THE GENIUS, AND THE THEORY OF RELATIVITY

WALTER ISAACSON

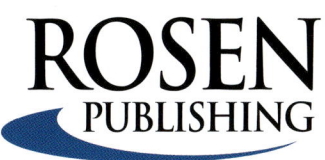

Published in 2022 by The Rosen Publishing Group, Inc.
29 East 21st Street, New York, NY 10010

Published in 2018 by Andre Deutsch, a division of the Carlton Publishing Group.
Design copyright © Carlton Books 2018.
Text copyright © Walter Isaacson 2009.

All rights reserved. No part of this book may be reproduced in any form without permission in writing from the publisher, except by a reviewer.

Cataloging-in-Publication Data

Names: Isaacson, Walter
Title: Albert Einstein / Walter Isaacson.
Description: New York : Rosen YA, 2022. | Series: Pioneers of science
Identifiers: ISBN 9781499471069 (pbk.) | ISBN 9781499471076 (library bound) | ISBN 9781499471083 (ebook)
Subjects: LCSH: Einstein, Albert, 1879-1955--Juvenile literature. | Physicists--Germany--Biography--Juvenile literature.
Classification: LCC QC16.E5 I87 2022 | DDC 530.092 B--dc23

Manufactured in the United States of America

CPSIA Compliance Information: Batch #CSRYA22.
For Further Information contact Rosen Publishing, New York, New York at 1-800-237-9932.

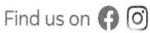

Contents

EARLY YEARS

Where Science Stood	4
Birth and Childhood	8
School	14

SWISS YEARS

Aarau	18
The Zurich Polytechnic	22
Mileva Marić	26
Lieserl	30
Patent Clerk	34
The Miracle Year: Quantum Theory	40
The Miracle Year: Special Relativity	46
The Rising Professor	52
Elsa Einstein	56

BERLIN YEARS

General Relativity	62
The Home Front	70
Divorce and Remarriage	74
The Eclipse	78
Einstein in America	86
The Nobel Prize	92
Quantum Mechanics	96
Einstein and Religion	102
The Rise of Hitler	106

PRINCETON YEARS

To America	110
The Bomb	116
Arms Control	122
Civil Rights	126
The Endless Quest	132
Israel	136
Red Scare	140
Farewell	148

Translations	156
Index	158
Credits	160

Where Science Stood

B y the end of the 19th century, the foundations of physics seemed to be firmly in place. Galileo had combined experimental observations with mathematical descriptions to construct a mechanical view of the universe. Isaac Newton had built on the discoveries of Galileo and others to come up with laws of motion and gravity that described a universe that was, at least in theory, fully predictable (a model which has come to be known as "classical mechanics"). Causes produced effects, forces acted upon objects, and a falling apple and an orbiting moon were governed by the same rules.

Newtonian mechanics was joined by another great advance in the middle of the 19th century: the discovery of the laws of electrical and magnetic fields. Michael Faraday, whose lack of formal education and unpromising background as the son of a blacksmith made his achievements all the more extraordinary, demonstrated that magnetism could be produced from an electric current, and that such an electric current could be created by the motion of a magnetic field. These realizations were carried even further by James Clerk Maxwell, a Scottish physicist who worked on the relationship between electrical and magnetic fields.

The electromagnetic field theory developed by Maxwell seemed, at least at first, to be compatible with the mechanics of Newton. Electromagnetic waves, such as light waves, were thought to be just one

ABOVE: *Galileo Galilei played a vital part in the scientific revolution that began in the Middle Ages and continued through the Renaissance. Albert Einstein referred to Galileo as the progenitor of modern science.*

THE THEORIES OF ISAAC NEWTON

Isaac Newton (1643–1723), a Cambridge mathematics professor, finally found answers to many of the problems that had vexed "mechanical philosophers" such as Robert Boyle and Robert Hooke and overturned a prevailing scientific view of the universe that still owed much to the thinking of the ancient Greek philosopher Aristotle. His theory of gravitational attraction provided a mechanical explanation of the orbits of celestial bodies in the solar system. His theories of motion, such as the law of inertia—that a body preserves its state of rest unless acted upon by a force—provided compellingly simple explanations of many physical phenomena, while his development of calculus (which he called "fluxions") provided a powerful tool for calculations involving curves and tangents.

ABOVE: *The title page of Newton's 1867 Philosophiae Naturalis Principia Mathematica.*

more phenomenon which fitted in with the framework of classical mechanics. Scientists simply assumed that the electromagnetic waves were caused by the vibrations and undulations of some form of omnipresent physical matter. They called this unseen substance the "light-bearing ether," and they assumed that it played a role in propagating light waves that was comparable to the role that water played for ocean waves and air fulfilled for sound waves.

With all these theories in place, the British physicist Lord Kelvin made a famous observation in 1900. "There is nothing new to be discovered in physics now," he told the British Association for the Advancement of Science. "All that remains is more and more precise measurement." We can sympathize with him. The combination of experimental observations and mathematical analysis exemplified by Galileo, Newton, and Maxwell had triumphed. The universe seemed governed by laws, and those laws seemed to be expressed by the language of mathematics.

But just at that moment, fissures were beginning to appear in the foundations of physics. Unexpected forms of radiation, such as x-rays and spontaneous radioactivity, were being discovered. The study of radiation, which happened when electromagnetic waves interacted with physical objects, showed that mysterious things were occurring at the intersection where Newtonian mechanical theories, which described discrete particles, ran into Maxwell's field theories, which dealt with electromagnetic phenomena. In addition, scientists had devised all sorts of ingenious ways to find evidence of the supposed light-propagating ether—but they had repeatedly come up empty handed.

In the year 1905, another great scientist, Albert Einstein, enters the story. At that point he was merely a third-class examiner in the Swiss Patent Office. He had graduated with mediocre grades from a teacher-training polytechnic in Zurich, where he had alienated most of his professors with his

"There is nothing new to be discovered in physics now" — Lord Kelvin

willingness to challenge authority. As a result, he had not been able to earn a doctorate or get a teaching job. But, because he was good at challenging assumptions and questioning premises, he had turned out to be a pretty competent patent examiner.

It was these traits that also helped to make him the right person to upend classical physics. He was a rebellious thinker, well suited to a moment when science needed to scrub away the layers of conventional wisdom that were obscuring the cracks in the foundation of physics. He was also very imaginative, which allowed him to make conceptual leaps that eluded more traditional minds. Above all, he was irreverent, so he could question assumptions that most scientists did not even notice they were making.

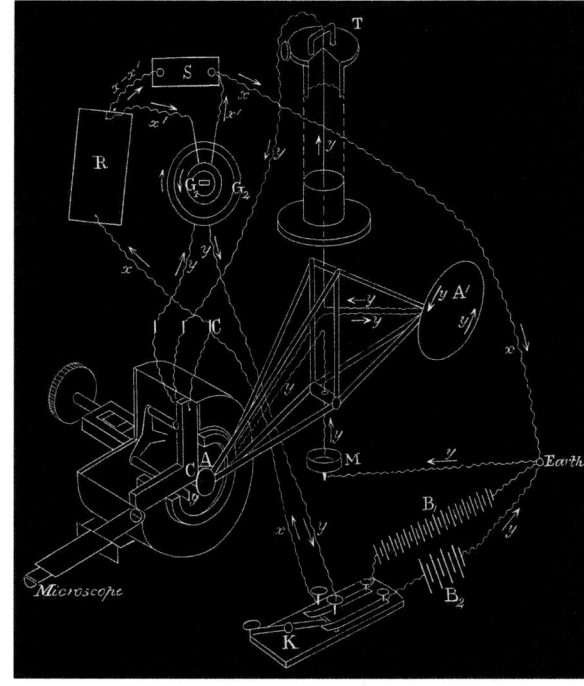

BELOW: *The apparatus Maxwell developed for the comparison of electrostatic and electromagnetic units.*

JAMES CLERK MAXWELL
(1831–1879)

Born in Edinburgh, James Clerk Maxell showed a very early talent for physics and studied at Edinburgh and Cambridge Universities, graduating from the latter in 1854. His early work included demonstrating that white light was composed of red, green, and blue light and explaining how Saturn's rings were able to remain stable. In the 1860s, he turned his attention to the relationship between electrical and magnetic fields, showing how changes in an electrical field could cause changes in a magnetic field, which in turn would provoke modifications in the magnetic field, producing an "electromagnetic wave." His paper "General Equations of the Electromagnetic Field" contributed vitally to the understanding of electric currents, conduction, and magnetism, and provided a theoretical framework for the ultimate construction of efficient electric motors.

Birth and Childhood

"the dopey one"
—the family maid on Einstein

Worried parents and underachieving students should take heart: Albert Einstein was no Einstein when he was a child. He was slow in learning how to talk, so slow that his parents consulted a doctor and the family maid dubbed him "the dopey one." His independent nature and defiance of authority led one schoolteacher to declare that the young Einstein would never amount to much.

This refusal to adopt a conventional approach would, so Einstein later surmised, contribute to his intense scientific creativity. And his slow verbal development led to his lifelong habit of thinking in pictures rather than just in words. He loved to perform what he referred to as visual thought experiments—what you and I might call daydreaming, but being Einstein, he gets to call them thought experiments.

Most of his great scientific breakthroughs would spring from just such imaginary experiments. What would a light wave look like if you were moving alongside it at the same speed? If two lightning strikes appeared simultaneous to a person on a train platform, would they appear that way to another person who was on a train that was speeding by? If you were in an enclosed elevator car accelerating upward in outer space where there was no gravity, would your experience be equivalent to that of being in an enclosed elevator car that was sitting on the ground in the earth's gravitational field?

His slow learning also caused him to marvel at the everyday phenomena that the rest of us take for granted. An example of this came when he was about five years old and his father gave him a compass. Einstein, who was home sick from school, became so excited when he examined it that his hands started to tremble. Nothing was touching the needle, yet it always pointed north no matter how you moved the compass. There was an invisible force field that pervaded his bedroom and, apparently, the entire universe.

Einstein's family background was modest; his ancestors were Jewish tradesmen and peddlers from rural Swabia who had achieved prosperity and become increasingly assimilated into German society and culture. He was born in March 1879 which

LEFT: *Einstein's Ulm birthplace.*

ABOVE: *Hermann Einstein, Albert Einstein's father.*

prophetically boasted as its motto, "*Ulmenses sunt mathematici*," ("the people of Ulm are mathematicians"). Ulm had at the time just become a part of the infant German Reich, along with the rest of Swabia. Einstein's parents, Pauline and Hermann, had originally planned to give him the name Abraham, for his paternal grandfather, but they changed their mind, considering their initial choice "too Jewish" (according to a later account by Einstein) and compromised by retaining the initial "A," but naming him Albert.

As a young child, Einstein delighted in constructing complex buildings from a toy set or creating card houses as high as 14 stories. But his persistence and tenacity were balanced by the tantrums to which he was prone; his sister Maja's head was the target of repeated attacks with hard objects.

Throughout his life, Albert Einstein would retain his childlike capacity for awe and wonder. He believed, as he expressed it when writing to a friend later in life, that people such as he did not age, but would instead forever retain a childlike curiosity in the face of the universe's great mysteries. He would always marvel at the astonishing phenomena of nature that most grownups take for granted. What is a magnetic field? What is gravity? Why does that compass needle twitch and point north? He was also always trying to picture and imagine things. What would it be like to ride alongside a light beam? What does a gravitational field look like?

Above all, he was rebellious enough to question any received wisdom, no matter how obvious. A foolish faith in authority, he repeatedly declared, was the enemy of truth. Newton had bequeathed the modern age with certain premises, which he proclaimed at the outset of his *Principia*, such as the assumption that time marches along, second by second, inexorably, independent of our observations of it. These things may have been obvious to others. But they made Einstein ask: How could we possibly know that?

BELOW: *The earliest photograph of Einstein that is known to exist.*

EINSTEIN'S SISTER

Einstein's sister Maria—who was always known as "Maja"—was born two years after him. Despite some difficult moments in childhood—the young Albert repeatedly battered her head with hard objects—she became her brother's lifelong confidante and often his closest friend. She qualified as a teacher at Aarau in 1902, and then studied Romance languages in Berlin. By the 1930s, she had moved with her husband, Paul Winteler, to Florence, but Mussolini's anti-Semitic legislation led her to join her brother in the United States in 1938. She lived for the next 13 years in Princeton, and, after she suffered a stroke in 1948, Albert nursed her, and would read to her every evening, works ranging from *Don Quixote* to esoteric ancient Greek scientific tomes. At her death, in 1951, Einstein was distraught.

ABOVE: *Einstein and his sister Maja in 1884, when Einstein was five and Maja was three.*

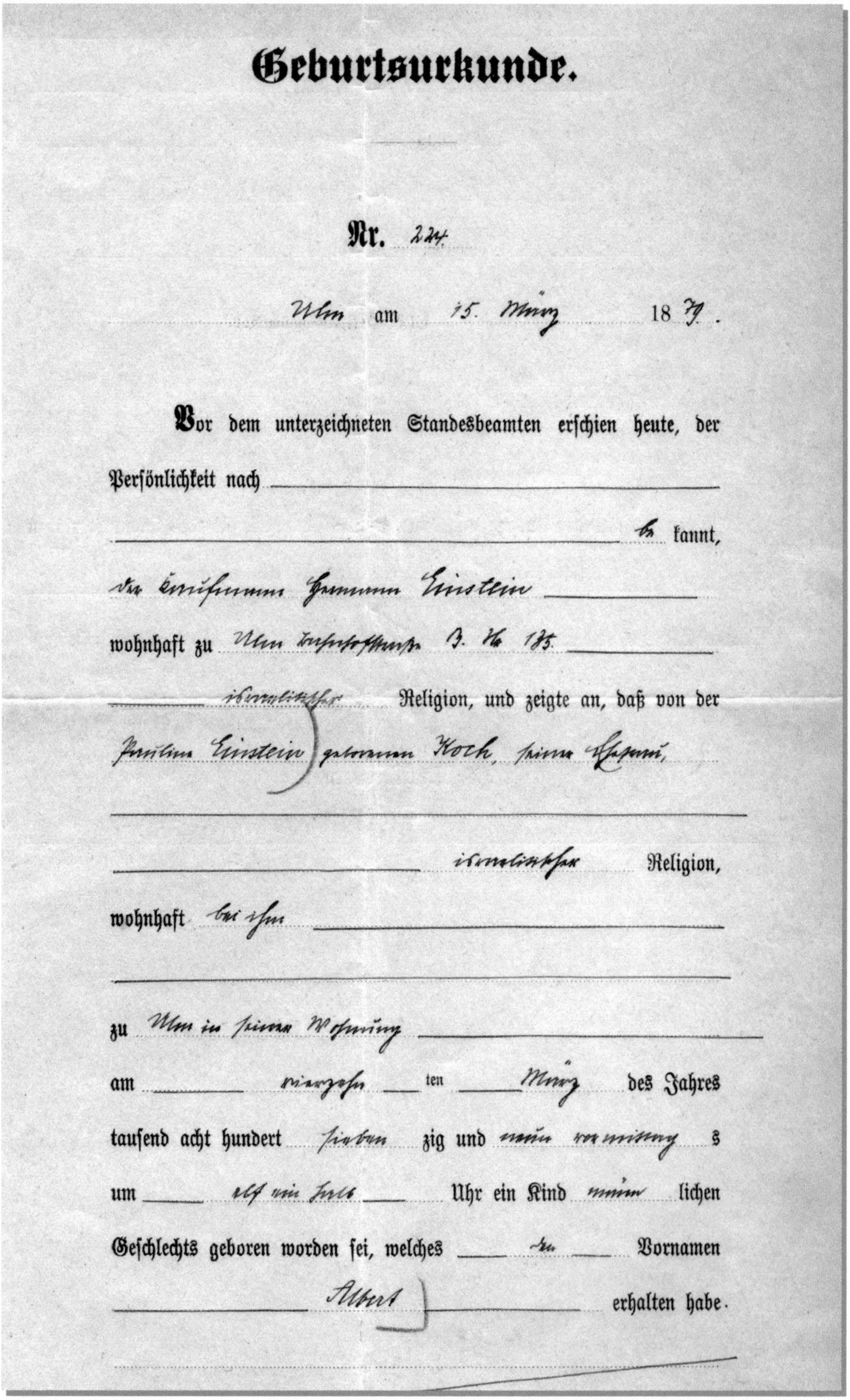

Geburtsurkunde.

Nr. 224

Ulm am 15. März 1879.

Vor dem unterzeichneten Standesbeamten erschien heute, der Persönlichkeit nach _____

_____ bekannt,

der Kaufmann Hermann Einstein _____

wohnhaft zu Ulm Bahnhofstraße B. No 135 _____

_____ israelitischer Religion, und zeigte an, daß von der

Pauline Einstein geborene Koch, seine Ehefrau, _____

_____

_____ israelitischer Religion,

wohnhaft bei ihm _____

_____

zu Ulm in seiner Wohnung _____

am _____ vierzehn _____ ten ____ März _____ des Jahres

tausend acht hundert _____ sieben _____ zig und _____ neun vorstehend _____ s

um _____ elf ein halb _____ Uhr ein Kind _____ männ _____ lichen

Geschlechts geboren worden sei, welches _____ den _____ Vornamen

_____ Albert _____ erhalten habe.

RIGHT: *Einstein's mother, Pauline.*

LEFT: *Einstein's birth certificate dated March 15, 1879.*

EINSTEIN AND MUSIC

When he was six, Einstein's mother gave him a gift — violin lessons — that would, like the compass, resonate throughout his life. Einstein rebelled against the mechanical discipline his music teacher tried to impose but after being exposed to Mozart's sonatas, he suddenly grasped the imaginative, creative spirit that inspires great music. He later told a friend that Mozart's music was so pure and beautiful that it seemed a reflection of the inner beauty of the universe itself. Music helped him to think. It was also his connection to the harmony of the spheres. Most importantly, it was a reminder that great genius, even (or perhaps especially) in the fields of mathematics and science, is a function not just of intelligence, but also of creativity and imagination.

RIGHT: *Einstein's love of music continued throughout his life and he liked to play his violin as often as possible.*

School

One delightful myth about Einstein is that he failed mathematics as a student in Munich. This assertion is so widespread in print and on websites, and so often accompanied by the confident phrase "as everyone knows," that it has virtually achieved the status of a received fact.

Alas, though Einstein's life was rich in delightful ironies, a lack of early mathematical ability is not one of them. He did not excel in languages, and his willingness to question authority meant he was not always the teachers' favorite. But he did well in mathematics, because he could visualize the concepts well. He thought that a mathematical equation was just God's brushstroke for painting something in nature, something that the imagination could see—just as you can conjure up a picture when you read Homer's phrase "rosy-fingered dawn." Einstein could visualize how equations were reflected in realities—how the electromagnetic field equations discovered by James Clerk Maxwell, for example, would manifest themselves to a boy riding alongside a light beam. He would always firmly believe that it was imagination that yielded the sharpest insights into underlying realities, and not plain knowledge of the facts.

His uncle Jakob Einstein, an engineer, instilled in him a love of algebra. His

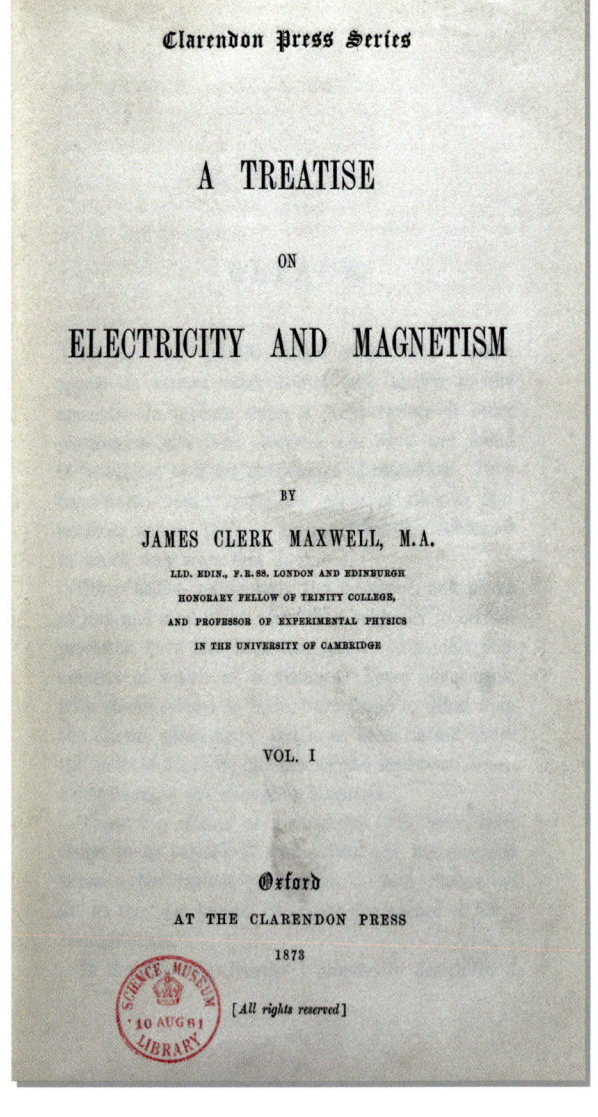

RIGHT: *Maxwell's most important work,* A Treatise on Electricity and Magnetism, *published in 1873, had a huge influence on the young Einstein.*

BERNSTEIN'S THOUGHT EXPERIMENT

Reading Aaron Bernstein's scientific books instilled in the young Einstein an interest in thought experiments. One topic that Bernstein dwelled on was the speed of light. In Bernstein's vivid treatment of the topic, we can see the germ of the thought experiments that Einstein would use 15 years later in wrestling with the theory of special relativity. Imagining what it would be like to be on a speeding train, Bernstein pointed out that a bullet through the train would appear to move at an angle, since the train would have moved between the bullet entering through one window and leaving through the other. The same must be true of light entering a telescope, because of Earth's movement through space.

BELOW: *An illustration of Bernstein's thought experiment. Diagram A shows the point where the bullet enters the train and diagram B shows the bullet's actual trajectory (the red line) and its perceived trajectory (green line.)*

approach, too, was imaginative. He likened the process of solving an equation to a hunt, in which the term "X" was the quarry to be tracked down ruthlessly until captured (or, more mathematically, solved). When Einstein mastered that, his uncle introduced some more difficult concepts. Among them was the Pythagorean theorem (the square of the length of each leg of a right-angled triangle add up to the square of the length of the hypotenuse). Eventually Einstein "proved" this theory using as his working method the similarity of triangles. He felt it was clear that the relationship between the sides of the right-angled triangles was wholly determined by one of the acute angles.

Once again, he was not merely memorizing concepts. Instead, he was visualizing them and thinking in pictures. It is not something that German schools emphasized when it came to mathematics, nor, for that matter, do most contemporary schools. The Pythagorean theorem tends to

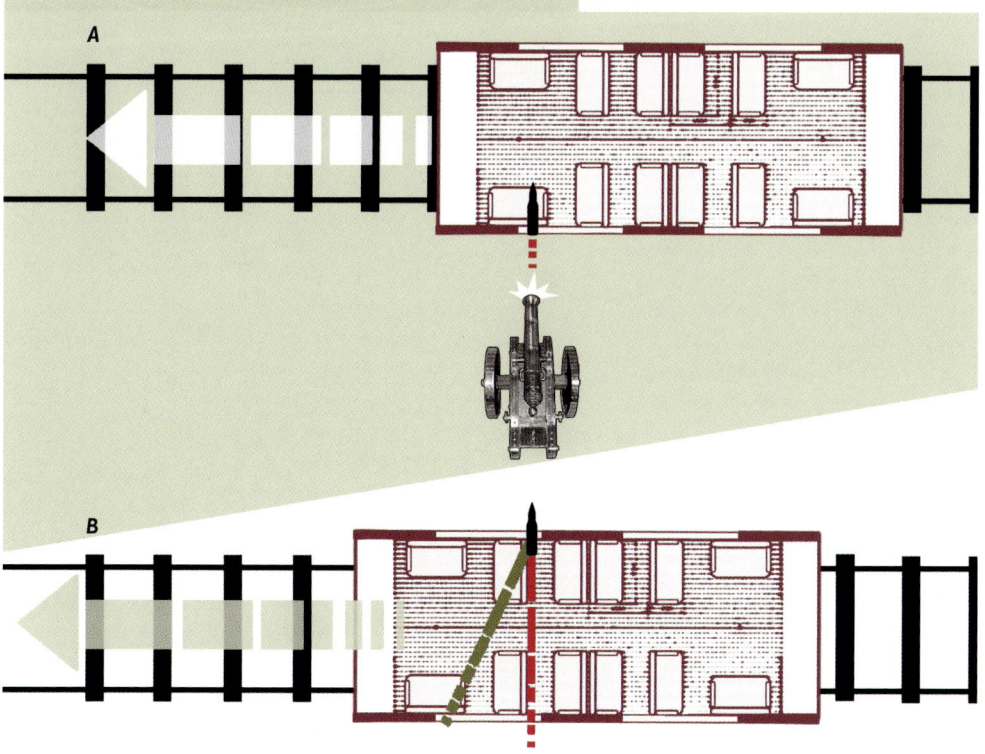

EINSTEIN AND THE MILITARY

The young Einstein would often see soldiers marching by his window to the accompaniment of military bands. Many of his schoolmates loved to play at being good soldiers and would pretend to march after the troops. But the whole spectacle caused Einstein to weep, and he told his parents that it was most distasteful. He did not want to grow up infected by such militarist notions; those who accepted them unquestioningly he regarded as little better than animals. This anti-authoritarian attitude caused him difficulties at school, as he found the style of teaching—rote drills and impatience with questioning—repugnant.

RIGHT: *Military parades such as this one, held in honor of the German crown prince enlisting in the Guard Regiment in 1896, greatly distressed Einstein when he was growing up.*

be taught by rote, rather than as something to be understood by picturing it in the mind's eye. His feel for the Pythagorean theorem would turn out to be useful as he visualized what would become his theory of special relativity.

Einstein had another great educational influence outside of school: a poor medical student named Max Talmud who came to dine with the Einstein family once a week. It was a long-established Jewish practice to invite a hard-up religious scholar to partake in their Sabbath meal. The Einsteins still held to this tradition, but in their version the date shifted to Thursday and the invitee was instead a medical student.

Talmud began his weekly visits when Einstein was 10. He brought with him popular science books, among them a series called *People's Books on Natural Science* written by Aaron Bernstein, which the young Einstein devoured with great attention, his breath almost taken away by the scientific insights it imparted. Its 21 small-format volumes contained a treasure chest of contemporary scientific work, especially that being carried out in Germany.

After Einstein had mastered Bernstein's books, Talmud gave him a geometry textbook so that he could revel in its joys before he was force-fed the subject in school. It was a book of which Einstein would later speak with almost reverential awe. It contained propositions of astonishing elegance, proved using the absolute certainty of geometrical axioms. That the three altitudes of a triangle intersected at a single point was by no means obvious, but the simple principles elucidated in Bernstein's work could provide a key to understanding such apparent complexities.

Even though his parents were secular and did not belong to a synagogue, Einstein spent a few years as a young boy embracing Judaism and even trying to keep kosher dietary laws. But his exposure to science produced a sudden transformation at age 12, just as he would have been readying for a bar mitzvah. Bernstein, in his popular science volumes, had reconciled science with religion. He achieved this by identifying a fundamental cause that lay behind the universe, whose discovery was the work of science, but the dim awareness of which represented

RIGHT: *A portrait of Einstein taken at a studio in Munich when he was 14.*

the religious consciousness inherent in all humans. That is a pretty good description of what Einstein would believe as an adult. But as a rebellious 12-year-old, his leap away from faith was a radical one.

This turning away from religion reinforced Einstein's natural skepticism about received wisdom. He developed an extreme distrust of dogma and religious authority, which would be reflected in the nonconformist bent of his later political, social, and scientific beliefs. Later Einstein said that this experience inspired in him a suspicion of all kinds of authority—a feeling that never left him. He likened this disregard for the conventional way of doing things to a guardian angel, which would guide him to places more respectful minds would never reach.

Around that time, when Einstein was 15, the electrical supply company run by his father and uncle began losing money. His family moved to northern Italy, where it was hoped there would be better prospects for a small firm. Einstein, it was intended, would stay behind for the next three years in Munich, lodging with a distant relative, so that he could finish school.

That was not destined to happen. Einstein, already considered an insolent student, was soon being encouraged, or perhaps even forced, to leave school. He convinced his family doctor, who was Max Talmud's older brother, to give him a letter declaring that he was suffering from nervous exhaustion. With this as justification, he left his school at Christmas vacation in 1894, took a train to Italy, and informed his alarmed parents that he was never going back to Germany. He pledged instead to study in order to gain acceptance the next autumn to the Zurich Polytechnic (now known as the Eidgenössische Technische Hochschule, or the Swiss Federal Institute of Technology).

Aarau

ABOVE:
This photograph was taken at the wedding of Einstein's sister, Maja, to Paul Winteler in 1910.

That summer of 1895, when Einstein was 16, he wrote his first physics essay. It was about a topic destined to play a major role in his career: the supposed substance known as "ether." Scientists conceived of light as a wave, and thus believed that the universe was permeated with an invisible substance, which they called ether, that propagated light waves by a ripple effect analogous to the way that water propagates waves in the sea. At the time, scientists were trying all sorts of methods to detect this ether and measure the earth's motion relative to it. None had been successful, but that did not deter any of them—or for that matter the teenage Einstein—from believing that ether must exist.

Einstein's paper outlined a program of experiments that could help explain how ether would behave in a magnetic field. He sent it to his uncle Caesar Koch, a merchant in Belgium. With feigned humility, he highlighted his paper's imperfections, disparaging it as the naïve production of a rather inexperienced youth. He also explained to his uncle his intention to enroll at the Zurich Polytechnic, but noted that the institution's rules on admission age made this difficult. as he was two years short of the official age for entry.

A friend of the Einsteins interceded on his behalf with the Polytechnic's chancellor, who agreed to permit the "so-called child prodigy" to take the admission test. He passed the section on mathematics and science easily. However, he failed in French, literature, politics, and zoology. One of the physics professors, Heinrich Weber, recognized Einstein's ability and suggested that he stay in Zurich and audit his classes, but Einstein decided that he would be better served by a year at a school in the nearby village of Aarau.

There he found the antidote to rigid German education. Einstein later wrote that the method of schooling in Aarau made him appreciate the value of an education that imbued a sense of taking responsibility for one's actions rather than obedience to rules and authority. The school also encouraged *Gedankenexperiment*—"thought experiments"—and these exercises in visual imagination would become one of the keys to Einstein's success.

In Aarau, he came up with what would become one of his most famous thought experiments: he tried to picture what it would be like to ride alongside a light beam. Although he himself felt his early thought experiments were childish, he later told a friend that this one had a direct relation to his special theory of relativity. He realized that if someone could chase after a light wave at the same speed as light, the light wave would, in effect, be frozen in time. He then dismissed this as impossible.

Einstein's host family in Aarau was the Winteler family, who would long remain a powerful influence in his life. No one was surprised, and everyone seemed happy, when he became romantically attached to their daughter Marie late in 1895, a few months after he moved in. She was just turning 18, two years older than he, and was staying with her parents prior to taking up an appointment at a school in a neighboring village. The next April, visiting his own family in Italy during his spring break, Einstein penned his first known love letter to Marie. He wrote how happy she made him, and how he had pressed her letter to his heart, picturing her writing it, her hands gliding back and forth across the paper.

Einstein was by birth a German citizen, but not a comfortable one. He disliked Germany's militarist atmosphere and worried that if he remained a citizen there, he would be subject to military duty. So, with his father's permission, he filed papers renouncing his citizenship, and the release came through in January 1896.

No longer a citizen of any state, he also did not consider himself a member of any religion. In his official request to renounce German citizenship, he had proclaimed that he was without any religious denomination. When he was older, especially as anti-Semitism began to boil up in Germany, Einstein would reconnect with his Jewish identity, unlike many of his colleagues who tried to renounce their Judaism. Always a nonconformist, Einstein was contemptuous of Jews who sought to assimilate and curry favor with their tormentors, dismissing them as snails abandoning their shells. Nothing could change a person's fundamental Jewishness. For Einstein the affiliation was tribal, not a matter of religious tenets. Although, as he explained, he did not hold any beliefs that counted as Jewish faith, he gladly counted himself among the Jewish people.

JOHANN HEINRICH PESTALOZZI'S THEORIES OF EDUCATION

Sometimes fate creates a perfect fit. That is what happened when Einstein landed at the Aarau School. The teaching there was based on the methods developed by Johann Heinrich Pestalozzi in the early 19th century. In his 1801 book, *How Gertrude Teaches her Children*, this Swiss educational reformer outlined a philosophy which encouraged children to think for themselves. He believed that they should move from observation and experimentation to a deeper understanding that eschewed rote drill and memorization. The Aarau School instead relied on conducting imaginative thought experiments, which Einstein loved. Even mathematics was taught by the Pestalozzi method, which began with observing objects and built up to the using visual imagination to devise abstract concepts. This approaching of fundamental concepts through visualization became a key part in shaping Einstein's genius.

LEFT: *A wood engraving from 1882 showing Pestalozzi and his wife Anna implementing their teaching methods.*

THE WINTELERS

Papa Jost Winteler, who was a Greek and history teacher, was a liberal social democrat with an edgy honesty and an idealistic streak, characteristics that Einstein found most appealing. He reinforced Einstein's aversion to German militarism and to nationalism in general. He also helped to instill in his young protégé a set of beliefs about internationalism, pacifism, and democratic socialism, as well as hopes for world federalism, that would long inform Einstein's own thinking. Other members of the family also exerted a profound influence on Einstein. Winteler's wife Rosa became a surrogate mother to Einstein, while in their daughter Marie, Einstein found his first girlfriend. Einstein's sister Maja would later marry the Winterlers' son Paul.

RIGHT: *Einstein's close friend Michele Besso provided Einstein with a further link to the Winterlers when he married their daughter Anna.*

After his year in Aarau, Einstein graduated with the second-highest grades in his class and again took the Zurich Polytechnic entrance exams. He gained a top grade in mathematics, despite muddling up an "imaginary" number with an "irrational" one. He also got a top grade in physics, even though he took only 75 of the 120 minutes allotted for that part of the test. French was the one section he struggled with. But although his mastery of the language was poor, the essay contained personal insights that make it the most interesting part of his exams. He wrote that he expected he would become a mathematics or physics teacher, concentrating on the theoretical aspects of those subjects. It was the independence that a scientific career would offer that made it attractive to him.

LEFT: *Einstein in 1896 with his graduating class in Aarau. Einstein is seated on the bottom left.*

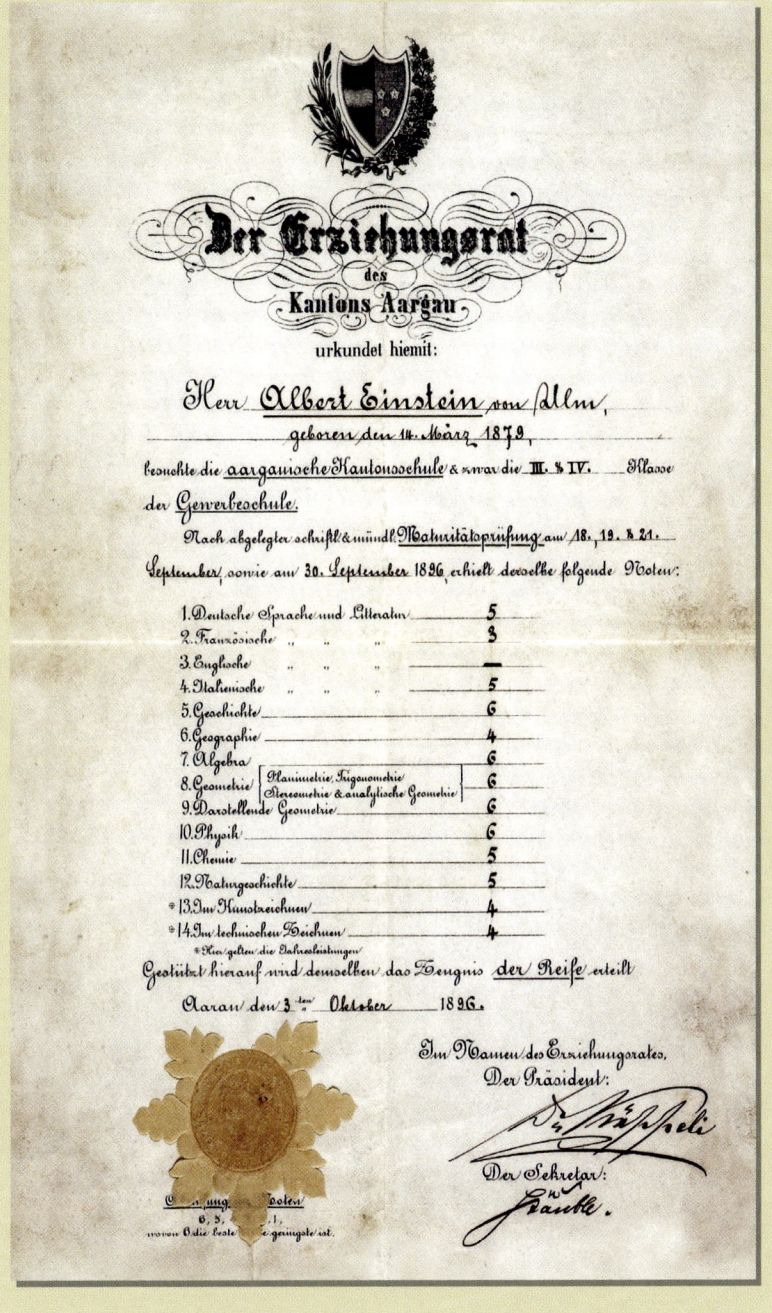

ABOVE: *Einstein's graduation certificate from Aarau in 1896. He received top marks in history, algebra, geometry, descriptive geometry, and physics. (See Translations, page 156.)*

The Zurich Polytechnic

The Zurich Polytechnic was the second-best college in Zurich. Unlike the University of Zurich, which overshadowed it, the Polytechnic did not grant doctoral degrees. It was primarily a teachers' and technical training institute. When the 17-year-old Albert Einstein arrived there in October 1896, he enrolled in the section that provided training "for specialized teachers in mathematics and physics."

Einstein was a good student—especially in physics. During the four years of his course, he got high marks in all of his theoretical physics courses, but lower grades in mathematics, and particularly in geometry. As Einstein later explained, he failed as a student to appreciate how a good grasp of the basic foundations of physics was a vital prerequisite to gaining mastery of more complex mathematical methods.

Even though he was a good student, he did not always get along well with his professors. A respect for authority, after all, was not part of his nature, and they did not fully appreciate the value of his rebellious and questioning personality. Heinrich Weber, who a year earlier had taken a liking to Einstein and urged him to audit his lectures, was his main physics teacher. They continued to get along well in Einstein's early years at the Polytechnic. But Einstein soon chafed at Weber's historical approach and lack of interest in the latest developments in physics. For example, Weber did not attempt any exploration of James Clerk Maxwell's elegant equations that described how electromagnetic waves such as light propagated. "We waited in vain for a presentation of Maxwell's theory," complained a fellow student. "Einstein above all was disappointed." He began to address Weber in an informal manner, calling him "Herr Weber" instead of "Herr Professor."

Einstein's disdain caused Weber to bristle. By the end of four years, they would be antagonists. "You're a very clever boy, Einstein," Weber advised him. "An extremely clever boy. But you have one great fault: you'll never let yourself be told anything." Whilst this was far from untrue, Einstein's ability to ignore received wisdom was not necessarily a fatal flaw, given the discordant state of the world of physics in the late 19th century.

> **"You're a very clever boy, Einstein. An extremely clever boy. But you have one great fault: you'll never let yourself be told anything."** —
> **Heinrich Weber**

ABOVE: *The Zurich Polytechnic, where Einstein was a student from 1896 to 1900.*

Einstein's relations with the other physics professor at the Polytechnic, Jean Pernet, were scarcely less difficult. Pernet supervised the practical experiments, and Einstein was never a great experimentalist, which is one reason he became a theorist. In Pernet's course "Physical experiments for beginners," Einstein got the lowest grade possible, a one. The professor thus gained the distinction of having failed Einstein in a physics course. His irregular attendance was in part to blame, resulting in March 1899 in a "director's reprimand" due to "lack of diligence in physics *practicum*."

When Einstein did actually turn up to Pernet's class, his stubbornly independent behavior often made matters worse. One day he was given written instructions for an experiment. "With his usual independence," a fellow-student later related, "Einstein naturally flung the paper in the waste paper basket." Refusing to acknowledge the instructions of others, he went about the experiments just as it pleased him. Finally, his waywardness caused an accident: in July 1899 Einstein injured his right hand in an explosion in the lab, and he had to have the wound stitched up at the clinic.

Music never ceased to be one of Einstein's ruling passions. It provided a direct connection with a sense of the harmony that lay behind the universe. The great composers' genius, he felt, lay in their ability to create order out of more than just words. The beauty of this harmony he also felt in his study of physics.

Sitting in his boarding house one evening, Einstein heard a Mozart piano sonata that was being played by an elderly female neighbor who gave piano lessons in the next-door attic room. He grabbed his violin, rushed out, and burst up the stairs of the neighboring house. Despite her bewilderment, Einstein implored her to go on playing, and within minutes the pair were playing a duet, with a violin accompanying Mozart's sonata.

Einstein especially appreciated the clarity of structure in the music of Mozart and Bach. He felt that it made their music appear to derive directly from the universe rather

MARCEL GROSSMANN (1878–1936)

When he studied, Einstein preferred the company of one or two close friends. Marcel Grossmann was the closest of these. From a prosperous Jewish background—his father was the proprietor of a factory near Zurich—Grossmann provided a useful source of lecture notes for Einstein, who was less than fastidious about even turning up. Einstein later acknowledged his debt in this regard to Grossmann, saying he preferred not to contemplate what would have happened to his grades had he not had his friend's notebooks to refer to. The pair would spend long afternoons drinking iced coffee and smoking at Zurich's riverside Café Metropole. "This Einstein will one day be a great man," Grossmann assured his parents. Grossmann himself would contribute to Einstein's future greatness by helping secure Einstein his first job at the Swiss Patent Office and by assisting him with the mathematics that converted the special theory of relativity into the general theory.

than being consciously composed, rather like his own scientific theories. Whereas Beethoven's music clearly came from a creative act on the part of the composer, Einstein later observed, that Mozart's music seemed to possess a purity that made it at one with the universe, coexistent and uncreated. Beethoven was little to his taste, and the fact that the music revealed so much about its creator left him feeling ill at ease.

At the Polytechnic, Einstein began to affect the rather distracted air and untidy appearance that would later make him the stereotype of an absent-minded professor. He often forgot to bring clothes with him when he traveled, and his persistent failure to remember what he had done with his keys became a standing joke with his landlady. On a visit to some family friends, Einstein had forgotten his suitcase and his host remarked to his parents: "That man will never amount to anything because he can't remember anything."

Now that Einstein was a typical student—in affecting an insouciant lifestyle and acting in a rather self-absorbed fashion—his relationship with Marie Winteler was unlikely to last. At first he kept up a strange correspondence with the sweet, but flighty daughter of his hosts in Aarau, sending her his laundry in the post, often without so much as even appending a note. Marie still sought to humor him, writing of her journey to the post office to return his freshly laundered clothes, of "crossing the woods in the pouring rain. In vain did I strain my eyes for a little note, but the mere sight of your dear handwriting in the address was enough to make me happy."

When Einstein ended the relationship with Marie, he justified himself in a letter to Mama Winteler. The letter shows how already Einstein had begun to shy away from the consequences of emotional commitments and sought science as a way of escaping from "merely personal" distractions. He justified his actions by maintaining that he shared some of the pain which he had inflicted on Marie by the abrupt nature of their break-up. Yet, self-absorbed, he went on to explain that he would now trust to intellectual pursuits and a contemplation of God's operation in the universe as the means to pull himself through these difficult times.

LEFT: *This 1895 picture shows Einstein's disheveled appearance. This led him to become the stereotype of the absentminded professor.*

Mileva Marić

LEFT: *Mileva Marić, the Serbian physicist and mathematician who would become Albert Einstein's first wife.*

There was only one woman in Einstein's mathematics and physics section at the Zurich Polytechnic. This was back in the days before girls were encouraged to go into the hard sciences, but Mileva Marić was an exception. She came from a humble Serbian peasant background, although her devoted father, who had pursued a military career and married well, had sufficient resources to support his talented daughter in her resolve to break into the male-dominated world of science.

Marić's education was a demanding one, although she always ended up top of her class in whichever institution she attended, including the Classical Gymnasium in Zagreb, in which, although it was notionally an all-male school, her father had managed to have her enrolled. Having graduated there, too, with top marks in physics and mathematics, Marić was just short of 21 when she joined Einstein's class at the Zurich Polytechnic.

Marić was not particularly pretty or charming. She was born with a dislocated hip that caused her to limp, and she had an air of despondency about her. But Einstein was attracted to her. She had qualities that he regarded, at least as an undergraduate, as enticing: a passion for science and a brooding soul. Her eyes had a depth and intensity about them. Her face had an alluring touch of melancholy. She would eventually become the keystone of Einstein's life, as his lover, wife, and sparring partner. The emotional field she created around him, and the force she exerted over him—both an attraction and at times a repulsion—was so powerful that even a scientist accustomed to unravelling ostensibly far deeper mysteries could never comprehend it.

They began to fall in love when they went hiking together in the summer of 1897, less than a year after they first met at the Polytechnic. Scared by the new feelings she was experiencing for Einstein, Marić decided to put some distance between them and enrolled briefly in classes at the University of Heidelberg.

Shortly after her arrival there Marić wrote Einstein a letter which demonstrates just the qualities in her that Einstein must have found so attractive. Far from the sweetness and flightiness of Marie Winteler, Marić exhibited a teasing detachment, pointedly observing that she had thought little about him, despite the long letter he had sent her. "It's now been quite a while since I received your letter," she said, "and I would have replied immediately and thanked you for the sacrifice of writing four long pages, would have also told of the joy you provided me through our trip together, but you said I should write to you someday when I happened to be bored. And

I am very obedient, and I waited and waited for boredom to set in; but so far my waiting has been in vain."

Einstein was apparently charmed by Marić's contradictory qualities: her levity and maturity, her nonchalance and focus, her passion and coolness. It was a mix that might seem peculiar, but it was also mirrored in Einstein's own personality. He tried to persuade Mileva to come back to the Polytechnic. By February 1898, much to Einstein's evident excitement, she had resolved to do so. It was not a decision she would regret, he wrote, urging her to come back as quickly as she could.

A few months later she was back in Zurich, living in a boarding house just a few blocks from Einstein. They became a couple, sharing books, intimacies, and access to each other's apartments. Their relationship puzzled Einstein's friends: his good looks and sensuous nature could have won him almost any woman he desired, yet he had chosen a rather unprepossessing Serbian with a bad leg and a melancholic nature. "I would never be brave enough to marry a woman unless she were absolutely healthy," a fellow student said to him. Einstein retorted that the beauty of her voice made Mileva hard to resist.

They were, indeed, kindred spirits, and their attraction was both an intellectual and spiritual one. They regarded themselves as an elite couple of academic outsiders. They rejected the norms of bourgeois society, and each sought romantic involvement with someone who could also answer their intellectual needs, and be a practical partner and collaborator. In his letters to Mileva, Einstein wrote that it seemed to him almost as if they were twin souls, taking delight in the same pleasures. It was commonplace shared experiences that bound them together—a hug, walking together, the making of coffee, even having a quarrel. And of course he relished her intellectual companionship and their time studying together. He took real pride in the idea that his girlfriend would soon have a PhD.

Their romantic and intellectual bonds were interwoven. While on vacation with his family in 1899, Einstein lamented in a letter to Marić that his first reading of Helmholtz was ruined

BELOW: *Pauline Einstein, circa 1900, the time when Einstein's relationship with Mileva was blossoming.*

EINSTEIN'S MOTHER AND MILEVA

After graduation, Einstein made his way in July 1900 to Melchtal, a Swiss Alpine village, where his family were spending the summer holidays. His mother had never met Marić, but she didn't like what she had heard about her. She immediately asked Einstein what would become of Marić. In his response, Einstein casually noted that she would eventually become his wife, so provoking a violent reaction from his mother who, as he later recalled, ran to her bed and cried inconsolably, her head buried beneath the pillow. It was not so much that Marić was not Jewish that caused Einstein's mother to reject her, for neither had been the much favored Marie Winteler. Just as were Einstein's friends, she was alienated by Marić's age—she was a few years Einstein's senior—her moody disposition, physical disability, and lack of good looks.

by not having her at his side. To work with her next to him was comforting and exciting.

When it came to the time to graduate, Einstein scraped by with a 4.9 average, beaten by three others in his small class of five. This was in marked contrast to his intermediate exams in October 1898, when he finished first in his class, with an average of 5.7 out of 6. Although the appealing myth that he failed mathematics at high school is totally unfounded, the nearly as pleasing story that he graduated almost at the bottom of his class is quite true.

Although it was a close thing, Einstein had done just enough to get his diploma, which he received in July 1900. Mileva Marić was not as fortunate. Her 4.0 mark was the only one which Einstein's 4.9 average bettered, and she was failed. Undaunted, however, she decided to stay on and try again the next year.

Einstein's feelings for her were undimmed by either her academic failure or the resistance of his family. His language at times grew colorful, referring to Marić as a wild street urchin, and declaring he was completely and hopelessly in love with her. In a poem he penned to her, he even imagined the flame of their passion had set their pillow ablaze.

MILEVA AT HEIDELBERG UNIVERSITY

Marić seems to have enjoyed her time at Heidelberg and found the lectures of assistant professor Philipp Lenard especially rewarding. That she found her time there intellectually stimulating is reflected in the intensity of her intellectual exchange with Einstein. She excitedly told him that Lenard was "talking now about the kinetic theory of heat and gases" and in another letter, she expounded on the ideas about the infinite which Einstein had expressed earlier, saying: "I do not believe that the structure of the human brain is to be blamed for the fact that man cannot grasp infinity. Man is very capable of imagining infinite happiness, and he should be able to grasp the infinity of space—I think that should be much easier."

ABOVE: *The German physicist Philipp Lenard. Mileva was impressed by his lectures at Heidelberg University.*

LEFT: *Heidelberg University. When Mileva became overwhelmed by the intensity of her relationship with Einstein, she briefly attended the university.*

ABOVE: *A postcard Mileva sent to Einstein early in their relationship. (See Translations, page 156.)*

Lieserl

In the spring of 1901, Einstein decided to ask Marić to go on vacation with him, even though he was not ready to defy his parents by marrying her. She had become despondent because of his parents' attitude, so Einstein suggested they spend a holiday in one of the world's most romantic spots: Lake Como, nestling jewel-like in the high Alpine lakes between Italy and Switzerland. During this vacation, Marić fell pregnant.

When she became pregnant by Einstein, Marić faced something of a dilemma. She was planning to retake her final exams at the Polytechnic, with the goal of attaining a doctorate and then becoming a physicist. She had worked to this end for so many years, and the prospect of motherhood threatened all of her emotional and financial investment. It would have been simple to put an end to the pregnancy: Zurich had become something of a center for the nascent birth control industry, including a mail-order firm which specialized in abortion drugs.

Yet she chose to go ahead and have Einstein's baby, even though the father was not as yet willing to become her husband. Having children outside marriage was unconventional at the time, if not exactly uncommon; such births made up 12 percent of the 1901 total number of births in Zurich. The figure was particularly high amongst

LEFT: *Einstein's first published scientific paper. It was on the capillary effect and was published in March 1901, around the time that Mileva became pregnant. When writing to Mileva about the paper Einstein often used the terms "we" and "our," which sparked a debate over how involved Mileva was with Einstein's scientific work.*

Austro-Hungarians, reaching the level of a third in southern Hungary, and even higher in Serbia. For Jews, however, the percentage was the lowest of all.

Einstein was now forced to look seriously at his own future. He was at that point only marginally employed, working as a part-time tutor for hire. Because he had alienated his professors, he had not been offered a fellowship to teach or pursue a doctoral degree, nor had he been able to get a favorable recommendation for any of the jobs he sought. Now that they were about to have a child, Einstein promised Marić that he would somehow find steady work, no matter how modest the position might be. No one, he added fiercely, would then be able to disparage her.

EINSTEIN AND MILEVA IN LAKE COMO

At the end of April 1901, Einstein summoned Marić to Como. He would brook no refusal, and commanded that she bring his blue dressing gown in which they would both wrap themselves up. He promised her an unforgettable trip, and on a Sunday morning in May, he eagerly waited for her at the train station there, his heart pounding. After a night in a local hotel, they hiked over into Switzerland through the mountains but found the pass blocked by snow. They completed their trek on a locally hired sleigh. In the coming days, Einstein would recall the closeness of their embrace as they pressed together, an act both beautiful and natural. It was in that most natural way that Albert Einstein fathered the daughter that he would never see.

BELOW: *A section of the shoreline in Lake Como, as it would have looked when Einstein and Mileva visited in 1901.*

LEFT: *Albert Einstein, circa 1900.*

Einstein's decision immeasurably improved the mood of Marić, who was having a difficult pregnancy in Zurich. "So, sweetheart, you want to look for a job immediately? And have me move in with you!" Einstein was not quite ready for that. To her evident dismay, he did not spend his summer with her, but again holidayed in the Alps with his mother and sister.

She moved to her parents' house in Novi Sad, where she gave birth in early 1902 to a daughter that she and Einstein called Lieserl. Einstein was not there. Instead, he was in Bern, where he was awaiting word on his job application to work as a clerk in the Swiss Patent Office. Einstein wrote to Marić wondering what the baby was like, and whether she was in good health. He speculated on which one of them the child most resembled, and how best her care might be arranged. Even though he had not yet seen Lieserl, he professed himself in love with her.

The baby brought out Einstein's wry side. Philosophically, he noted that while the baby certainly knew naturally how to cry, laughing was something she would have to learn rather later. Yet his love for their new baby seemed to exist mainly in the abstract, for it was not quite enough to induce him to make the train trip to Novi Sad.

Einstein kept Lieserl's birth a secret from his family and friends. It seems, indeed, that he may not ever have taken them into his confidence about his daughter. The baby girl was apparently put up for adoption, and as far as we know, Einstein never laid eyes on her. He never spoke about her in public or dropped any hint of her existence. He never referred to her in his correspondence, save in a few letters to Marić which were kept a secret until 1986, when Lieserl's existence was revealed to the totally unsuspecting academic world.

In those few surviving letters, the last mention of Lieserl comes two years later, in September 1903. It is likely that she died of scarlet fever around that time, but no one today knows for sure. We do know that she never came to Bern, where her father was able to retain his freedom untrammeled, and the aura of a respectable bourgeois that was necessary to further his quest to become an official of the Swiss government.

BELOW: *Einstein with his first wife Mileva in 1905. By this time, Mileva had given up her studies to concentrate on being a wife and mother.*

MILEVA'S STUDIES NEGLECTED

Mileva suffered from illnesses during her pregnancy which left her bedridden. This, and the worry caused by the forthcoming birth, must have made it difficult for her to study and Einstein hardly helped by his failure to come to Novi Sad to see her, despite her attempts to persuade him to come and seek her parents' approval. In late July 1901, she re-sat her exams, but once more she did not pass them. She got exactly the same mark, 4.0 out of 6, as at her first attempt, and once more she was the only one in her group who failed.

Patent Clerk

Einstein's life story holds many surprises, but amongst the greatest of all are the problems he encountered in landing himself an academic position. It was no less than nine years, an extraordinary interval, before he secured a junior professorship. And even more astonishing is that it was not until four years after he had revolutionized physics in his "*Annus Mirabilis*," or "Miracle Year," before anyone saw fit to offer him academic employment. None of his professors from the Polytechnic would give him a positive recommendation.

LEFT: *49 Kramgasse, Bern. This was Einstein's home while he worked at the Patent Office and developed his "Annus Mirabilis" papers.*

As he searched for a job, Einstein continued to write papers about physics, none apparently presentable enough to be accepted as a doctoral dissertation. He wrote the first of these on the capillary effect, the phenomenon exhibited when water curves upward as it clings to the side of a straw. Einstein later had a low estimation of this early work, dismissing the paper as valueless. It is, however, interesting for the development of his thought, as he explored an idea that would lie at the heart of his next five years' work—that atoms and molecules

DAVID HUME'S THEORIES

Einstein once said that the greatest influence on him was David Hume (1711–1776), the Scottish empiricist. His thinking was the culmination of a philosophical tradition that included John Locke and George Berkeley, and which eschewed reliance on any knowledge other than that which was derived directly from the senses. He even questioned the intuitive laws of causality, maintaining that just because an event (such as one ball hitting another and causing it to move) happened time after time, apparently without fail, this was no good reason to maintain that it would always do so. Hume applied his empiricism to the concept of time, denying it an existence that was independent of the observable changes in objects that permit us to measure its course. This assertion that absolute time has no meaning would resonate strongly in Einstein's later work.

RIGHT: *One of British painter David Martin's most influential works was his portrait of David Hume. Einstein said he had been inspired by Hume's views when he developed his special theory of relativity.*

Patent Clerk 35

do exist and that a variety of phenomena in nature can be understood by examining their statistical behavior.

Finally, Einstein got a break. Marcel Grossmann, the friend from college who took notes for him in mathematics class, had a family connection which could help get Einstein a job at the Swiss Patent Office in Bern. It would have to be as a third-class examiner, the lowest rank, because a doctorate was required to be a first or second-class examiner and the authorities in Zurich had rejected Einstein's repeated attempts to have a doctoral dissertation accepted.

In June 1902, the confirmation of Einstein's position finally came through. The Patent Office was in Bern's Postal and Telegraph Building, near where the city's famous clock towered over one of its medieval gates. When the clock struck each hour, a parade of figures would come out, with jesters, dancing bears, a knight in full armor, and finally, Father Time himself bearing an hourglass. At the nearby train station, the platforms were lined with stately clocks, each synchronized with the one in the tower. The trains arriving from distant cities would check their own timepieces with a quick glance at the clock on the tower as they sped through Bern.

It was in this most unlikely setting that Einstein spent the most productive seven years of his scientific life, arriving for work six days a week at 8 a.m. to examine applications, even after he had written the papers that transformed physics. He recounted to a friend the incredible business of his daily life, with eight-hour days spent at the Patent Office, followed by an hour of giving private lessons, and only then being able to turn to his own scientific work.

But we should not feel sorry for him. The job at the Patent Office was probably a godsend, better than being a junior professor at a university. He would not have made a good acolyte in the academy, for he would have recoiled at teaching the conventional dogma, at trying to please senior professors, and churning out safe publications. Instead, when looking at patent applications he

ABOVE: *The clock tower in Bern in 1900. It was this view that inspired Einstein two years later.*

got to do what he did best: question all of the premises, challenge any assumptions, and visualize how the underlying concepts would work in reality. Focusing on real-life technical and physical questions was a saviour for him intellectually. It meant that he understood better than many of his scientific contemporaries how theoretical concepts had real physical implications in practice.

Among the patent applications he considered were many for devices to synchronize distant clocks. The Swiss had adopted standard time zones and, being Swiss, wanted to make absolutely sure that when it struck seven in Bern it would strike seven at that exact same instant in Zurich and elsewhere. These applications had one thing in common: to synchronize two distant clocks, you have to send a signal between them, and the typical signal travels at the speed of light, whether it is a light or radio or electric signal. All the while, Einstein was still mulling over that thought experiment he had done at age 16, about what it would be like to catch up with a light wave.

Einstein had in Bern a group of friends who loved to discuss ideas. Among them were Conrad Habicht, who had studied mathematics at the Zurich Polytechnic, and the Romanian Maurice Solovine, who was a philosophy student at Bern. In a satirical take on serious academic societies, they called themselves the Olympia Academy. As the oldest, Einstein was made its president, and Solovine drafted a certificate with a profile portrait of Einstein crowned by a string of sausages.

The group's late night debates would often be rounded off by a violin performance from Einstein, or, in the summer, they would sometimes watch the sun rise from the top of a peak at the edge of Bern. "The sight of the twinkling stars made a strong impression on us and led to discussions of astronomy," Solovine recalled. "We would marvel at the sun as it came slowly toward the horizon and finally appeared in all of its splendor to bathe the Alps in a mystic rose." Once the café on its slopes opened up for business, they would down some strong coffee before trekking into the city for their day's work.

The reading list that the Academy set itself was a formidable one: it ranged from classics such as Sophocles's *Antigone*, whose theme of defying higher authority Einstein must have found sympathetic, to Miguel de Cervantes's *Don Quixote*, its windmill-tilting theme also having something appropriate to say about stubborn tenacity. Yet it was along the borders between science and philosophy that the three Olympians found their most fruitful texts: David Hume's *A Treatise of Human Nature*, Ernst Mach's *Analysis of the Sensations and Mechanics and its Development*, Baruch Spinoza's *Ethics*, and Henri Poincaré's *Science and Hypothesis*. It was from works such as these, and in particular those of David Hume and Ernst Mach, that Einstein derived much of his own scientific philosophy.

On January 6, 1906, Albert Einstein and Mileva Marić were finally married, in a civil ceremony at the Bern Register Office. Amongst the few guests were the members of the Olympia Academy, while no one from either bride or groom's family attended. After a celebration with their tight-knit group of intellectual friends, Albert and Mileva retired to their apartment. True to form, Albert had forgotten his key and had to wake up his landlady.

The couple's second child, a son named Hans Albert Einstein, was born in May 1904. Einstein was, at least initially, a good father, and built toys for the little boy, including a cable car he made out of string and matchboxes. "That was one of the nicest toys I had at the time and it worked," Hans Albert would remember when grown up. "Out of little string and matchboxes and so on, he could make the most beautiful things."

RIGHT: *Einstein and the members of the Olympia Academy. From left: Conrad Habicht, Maurice Solovine, and Albert Einstein. The photograph was taken around 1902.*

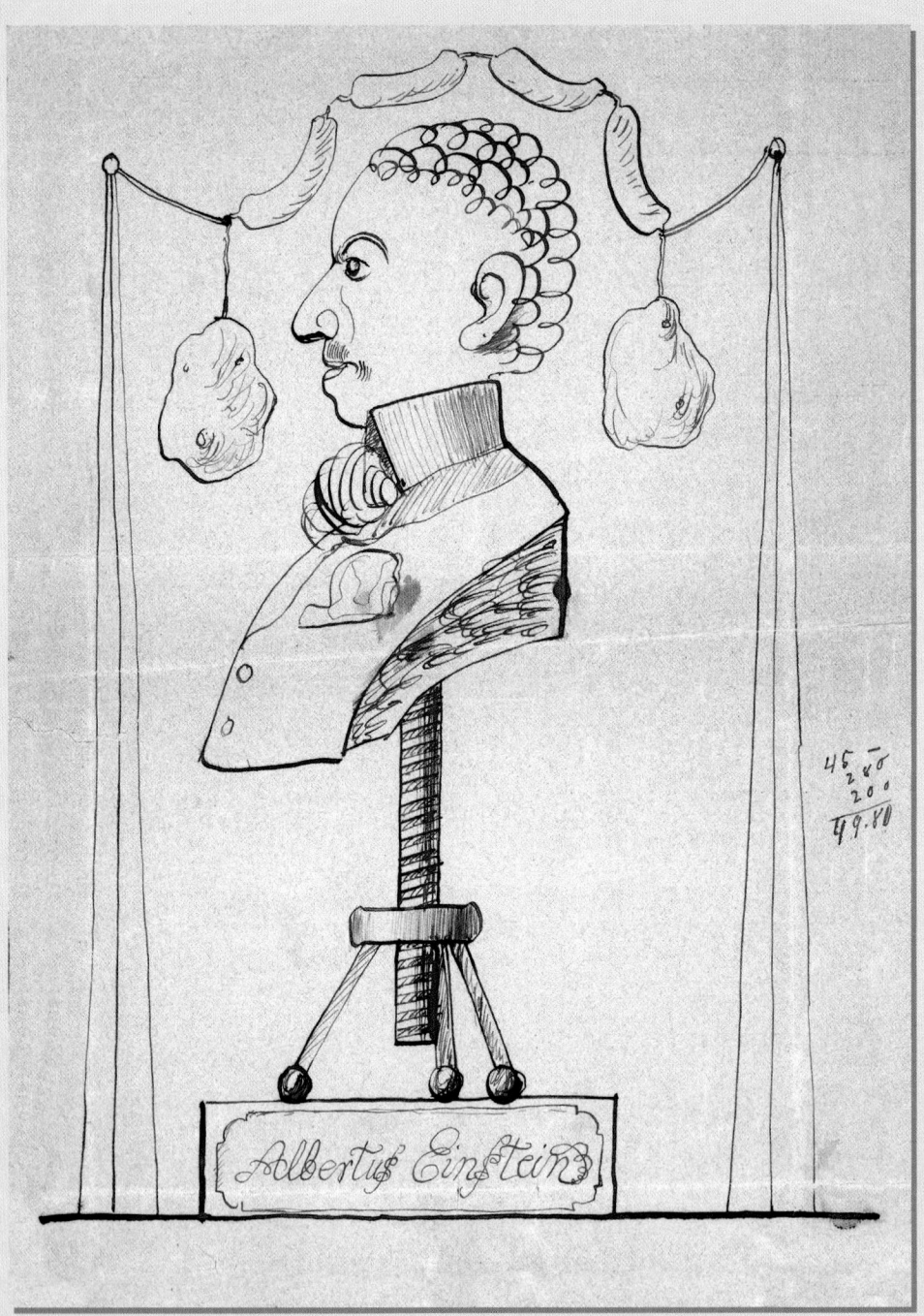

ABOVE: *The Olympia Academy certificate that was drawn for Einstein by Maurice Solovine.*

ERNST MACH'S THEORIES

The Austrian Ernst Mach carried Hume's empiricism even further and helped reinforce Einstein's skepticism about received wisdom. Einstein summarized Mach's philosophy as considering concepts meaningful only if they referred to concrete, observable objects and the rules by which these objects operated. In other words, for a concept to make sense you need an operational definition of it, one that describes how you would observe the concept in operation. Mach applied this approach to Newton's concepts of "absolute time" and "absolute space." Since these could not be defined by observable phenomena, they lacked meaning. This would bear fruit for Einstein when, a few years later, he would question what observation would give meaning to the apparently simple concept that two events happened "simultaneously."

LEFT: *A 1910 photograph of Ernst Mach. Although Einstein credited Mach and his theories as being precursors to the theory of relativity, Mach eventually denounced it.*

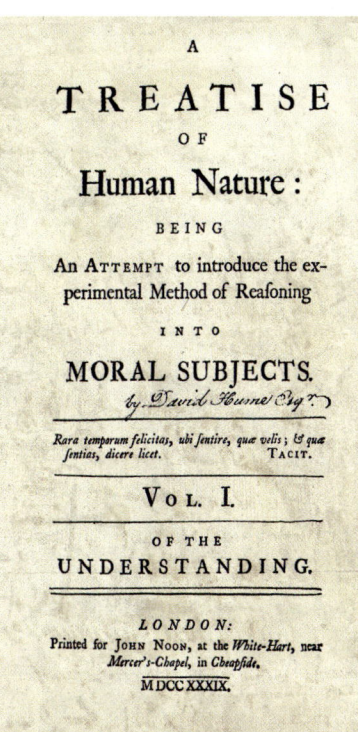

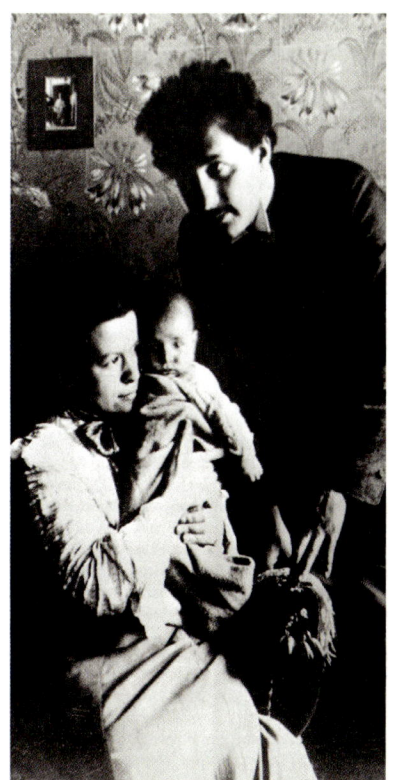

FAR LEFT: *The title page of the first edition of the first volume of David Hume's* A Treatise of Human Nature. *It was in this book that Hume outlined his theories of space and time that were to become so influential to Einstein.*

LEFT: *Mileva and Einstein at their home in Bern in 1904 with their newborn son Hans Albert.*

The Miracle Year: Quantum Theory

Conrad Habicht, Einstein's companion in the Olympia Academy, left Bern during the spring of 1905, and he proved a little lazy about corresponding. Happily for historians, this gave Einstein cause to write a letter chiding him. This letter would turn out to bear some of the most significant tidings in the history of science, a description of what has since become known as Einstein's "Miracle Year." But the letter's momentous nature was masked by an impish tone that was typical of its author.

Einstein apologized for writing a letter that was all prattle and nonsense but castigated Habicht for not sending a copy of his dissertation. Then he offered a deal: if Habicht would send his dissertation, in return Einstein would send him four papers. Only when Einstein got around to describing these papers, which he had produced during his spare time, did he give some indication that he sensed their significance.

The first of these dealt with the energy properties of light and radiation and Einstein felt that it was genuinely groundbreaking. It was indeed revolutionary. It argued that light could be regarded not just as a wave but also as a stream of tiny particles or packets called quanta. The implications that would eventually arise from this theory—a cosmos without strict causality or certainty—would spook Einstein for the rest of his life.

In his second paper, Einstein turned to how to determine the actual size of atoms. Even though the very existence of atoms was still in dispute, this was the most straightforward of the papers, which is why he chose it as the safest bet for his latest attempt at a doctoral thesis. He was in the process of revolutionizing physics, but he had been repeatedly thwarted in his efforts to win an academic job or even get a doctoral degree, which he hoped might get him

LEFT: *A computer illustration of subatomic particle tracks in white and yellow on a star field. The tracks are the result of particles moving through a bubble chamber containing superheated liquid. Every particle has a distinctive pattern, and particle physics has shown the basic structure of the atoms from which everything is made.*

ABOVE: *The contents page for the issue of* Annalen der Physik *(Annals of Physics) in which the first of Einstein's papers in his "Miracle Year" was published.*

promoted from a third to a second-class examiner at the Patent Office.

The third paper explained the jittery motion of microscopic particles in liquid by using a statistical analysis of random collisions. In the process, it established that atoms and molecules actually exist.

The final paper, Einstein wrote to Habicht, was still just a rough draft. It dealt with the electrodynamics of moving bodies, and involved a total reassessment of traditional notions of space and time. It was hardly the inconsequential rambling which Einstein labeled it. Based purely on thought experiments—performed in his head rather than in a lab—he had decided to discard Newton's concepts of absolute space and time. It would become known as the special theory of relativity.

What he did not tell his friend, because it had not yet occurred to him, was that he would produce a fifth paper that year, a short addendum to the fourth, which posited a relationship between energy and mass. Out of it would arise the best-known equation in all of physics: $E=mc^2$.

Einstein was right when he noted that the first of these 1905 papers would end up being regarded as revolutionary. Incorporated within it, indeed, was a notion that heralded a sea change in the development of physics. The idea that light is not made up solely of waves, but also consists of tiny packets (or quanta, which came later to be known as photons) transports physics into obscure realms that are more elusive even than the most apparently bizarre phenomena associated with the theory of relativity.

Einstein's papers addressed concerns which were vexing physicists around 1900—in fact they had done so ever since the ancient Greeks and have still not been entirely resolved. They concerned the composition of the universe, whether particles such as atoms and electrons make it up, or whether it could be better described in terms of an unbroken field, such as a gravitational or electromagnetic one. Furthermore, as both conceptions of the universe seemed equally valid at different times, what would happen in a situation where they intersected?

As well as seeking to explain oddities about light such as "blackbody radiation," scientists also sought to solve the puzzling phenomenon known as the "photoelectric effect." When a metal has light shone on it, the electrons on the surface are knocked off and emitted as radiation. Experiments

BELOW: *A blue flame from a Bunsen burner is used to heat an iron nail.*

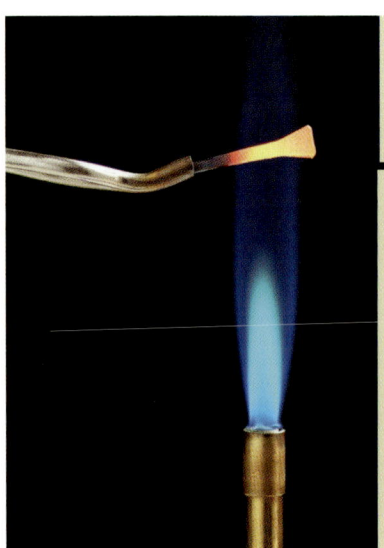

BLACKBODY RADIATION

Scientists in the late 19th century had been studying the intersection between waves and particles by looking at what was called "blackbody radiation." When metals such as iron are subjected to heat, the resulting glow changes color as the metal becomes hotter. Starting with red light, the glow becomes orange as the heat gets more intense, then changes to white, and finally, at the highest temperatures, appears blue. To try to understand the nature of the radiation, scientists devised a metal container. This was sealed, save for a tiny hole through which a small quantity of light could leak. This enabled scientists to measure the wavelength of light emitted at each new temperature level.

A WAVE

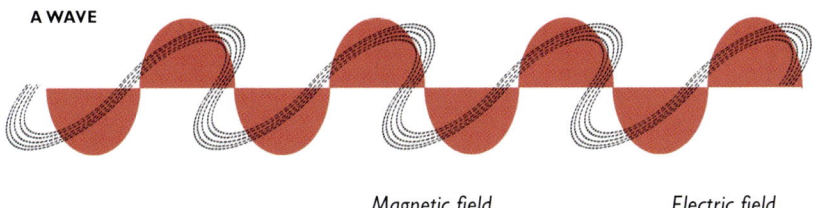

Magnetic field   *Electric field*

A STREAM OF PARTICLES

Photon

ABOVE: *An illustration showing the photons that make up a light wave.*

conducted by Philipp Lenard, later to be an anti-Semitic antagonist of Einstein, showed something unexpected. The wave theory of light suggested that when the intensity of the light increased, electrons with more energy and speed should be produced. The experiments showed, however, that although the greater the intensity of the light, the greater the number of electrons emitted, the energy of these electrons did not increase.

Einstein's triumph in his revolutionary 1905 paper on light quanta was that he made a leap of the imagination—thinking out of the box—that was in fact very obvious. He visualized the reality behind Max Planck's equation and the mathematical quirk that it contained. From that, he came to the conclusion that light really was made up of particles. In one revolutionary sentence in his 1905 light quanta paper, he determined that light was not continuously distributed in space but instead made up of a definite number of energy quanta, which have a definable location in space and which are produced by light sources or absorbed into objects as individual units.

Yet Einstein was keen to stress that his new insight did not mean the discarding of traditional wave theory, which dealt in terms of continuous spatial functions. This remained equally valid in a narrower sphere and worked particularly well in explaining optical phenomena. Einstein had fanned the flames of Planck's insights, producing a conflagration that would utterly overwhelm the world of classical physics. Planck had viewed the quantum as a mathematical notion, which helped the equations regarding the emission and absorption of light work more smoothly, and never saw them as something real and fundamental to the nature of light. In contrast, Einstein regarded the light quanta—which in 1926 came to be known as photons—as something inherent in the reality of the cosmos, no matter how bizarre, puzzling or confounding their existence might seem to be.

Einstein quickly realized that this quantum theory could undermine classical physics. He later wrote that his attempts to make the traditional theoretical physics conform to his new insights all failed. The foundations of the old certainties had been undermined, and he felt destabilized.

Einstein then related this new theory to the photoelectric effect. If light behaved like a particle, that would explain Lenard's experimental results. Einstein's theory produced a law of the photoelectric effect that was experimentally testable: the energy of emitted electrons would depend on the frequency of the light according to a simple

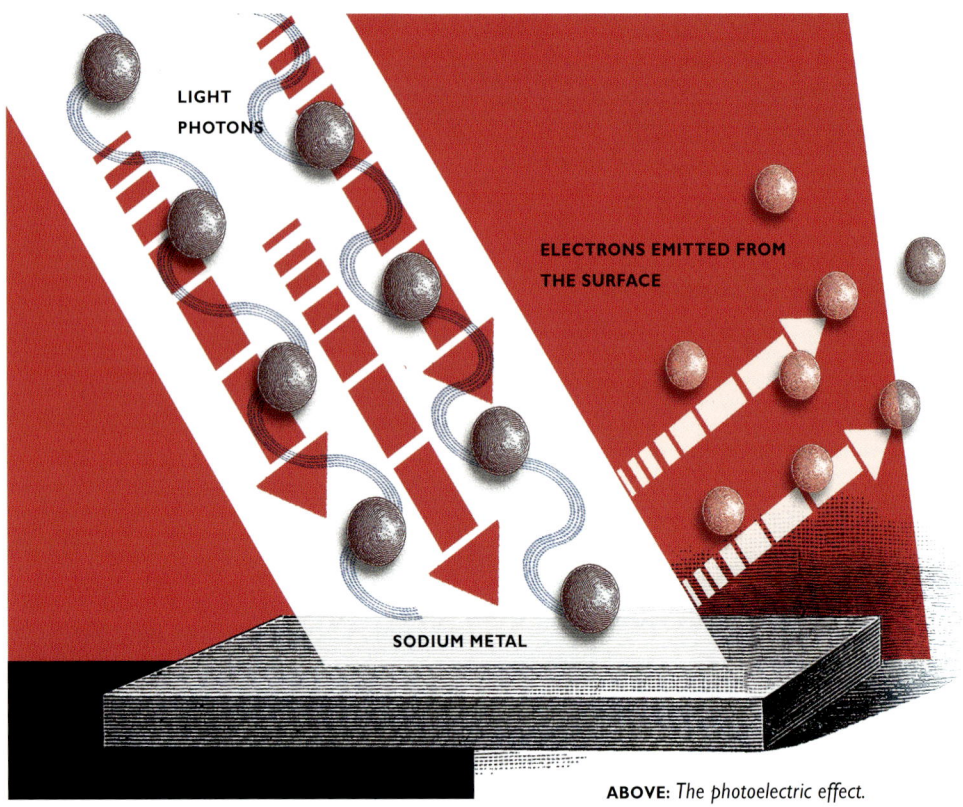

ABOVE: *The photoelectric effect.*

mathematical formula that involved Planck's constant, a calculation that was later verified. It would eventually (17 years later) turn out to be the one discovery made by Einstein that was uncontroversial enough to win him his only Nobel Prize.

Although Einstein had turned classical physics on its head, he still had not been awarded a doctorate. He now resolved to try once more to have a dissertation accepted. He selected a sufficiently safe subject matter, using a paper he was working on that found a new way to calculate the size of molecules. It satisfied the Zurich Polytechnic authorities who finally conceded the doctoral degree. As it happens, this was the paper of Einstein's which had perhaps the widest practical application, finding use in such areas as aerosols, the manufacture of dairy products, and cement mixing.

In this most productive phase of his life, it took Einstein a few weeks to come up with yet another paper analyzing the fundamental characteristics of atoms and molecules. Scientists had been struggling for eight decades to come up with an explanation for Brownian motion, the apparently random moving around of small particles in water or other liquids. Through his work on the subject, Einstein established definitively the physical existence of atoms and molecules.

Part of Einstein's genius was that he could turn his mind to fielding a number of radically different ideas at the same time. At the very time he contemplated the Brownian jigging about of molecules, he was concocting a theory with profound implications for motion and the speed of light. Within days of submitting his paper on Brownian motion, a conversation he was having with Michele Besso sparked a profound insight. He wrote soon after to Habicht saying his new idea amounted to a complete reshaping of traditional theories about space and time.

MAX PLANCK'S THEORIES

As they studied blackbody radiation, scientists noticed the graphs they came up with were shaped in the form of a hill. Their efforts to devise a mathematical formula that would fully account for this met with little success until, in 1900, the task was finally accomplished by the great German theorist Max Planck. His equation solved the problem by incorporating a constant, a small amount which needed to be added in to make the calculations work but which was otherwise unexplained. Planck believed his constant was a mere convenience to force the equations on light absorption and emission to work. He did not consider that it hinted at something fundamental in the nature of light itself.

RIGHT: *Although Planck had first put forward his solution in 1900, it was not until Einstein's 1905 paper on light quanta that the full implications of Planck's equation were understood.*

> "...a modification of the theory of space and time"
> —Michele Besso on Einstein's new theories

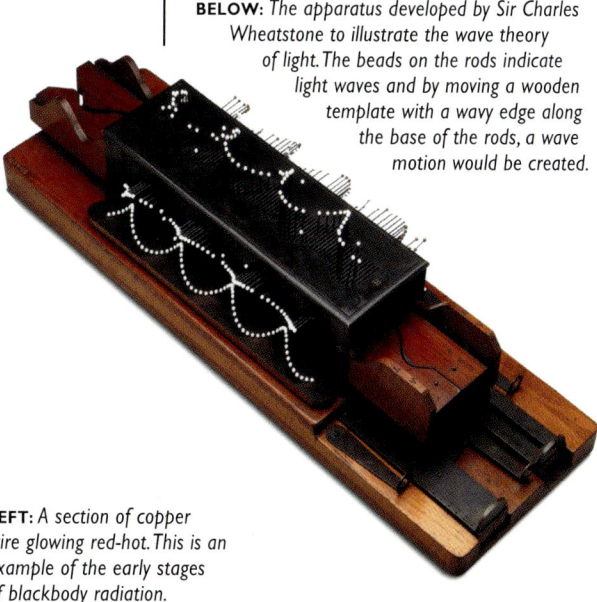

BELOW: *The apparatus developed by Sir Charles Wheatstone to illustrate the wave theory of light. The beads on the rods indicate light waves and by moving a wooden template with a wavy edge along the base of the rods, a wave motion would be created.*

LEFT: *A section of copper wire glowing red-hot. This is an example of the early stages of blackbody radiation.*

The Miracle Year: Special Relativity

In the 17th century, the critics of Copernicus and Galileo contended that Earth could not be moving, for if it were, we would feel it. Galileo countered their objections with a memorable thought experiment about a sailing ship:

"Shut yourself up with a friend in the main cabin below decks on a large ship, and have with you there some flies, butterflies, and other small flying animals. Have a large bowl of water with some fish in it; hang up a bottle that empties drop by drop into a wide vessel beneath it. With the ship standing still, observe carefully how the little animals fly with equal speed to all sides of the cabin. The fish swim indifferently in all directions; the drops fall into the vessel beneath; and, in throwing something to your friend, you need throw it no more strongly in one direction than another, the distances being equal; jumping with your feet together, you pass equal spaces in every direction. When you have observed all these things carefully, have the ship proceed with any speed you like, so long as the motion is uniform and not fluctuating this way and that. You will discover not the least change in all the effects named, nor could you tell from any of them whether the ship was moving or standing still."

Conversation inside Galileo's ship can also be conducted with ease, as the very air inside it moves at the same speed as the people inside it, and it is this air which conducts the sound wave. If someone were to drop a stone into a container of water on the ship, the ripples it produced would not differ from a similar bowl of water ashore, because, just like the air, the water is moving at the same time as the rest of the ship.

But what happens when you go outside on deck? The speed of a sound wave from a fellow passenger or a distant horn will depend on your motion relative to the air that is propagating the sound wave. Likewise, the speed that the waves in the ocean will come toward you depends on how fast you are moving through the water.

Ever since he was 16, this had led Einstein to imagine what it would be like if one could ride beside a beam of light. Would the light waves behave the same way as water or sound waves? Einstein's youthful thought experiment had been prompted by looking at the equations discovered by James Clerk

LEFT: *Copernicus, the first astronomer to develop the theory that the sun is the center of the solar system and the Earth orbited around it.*

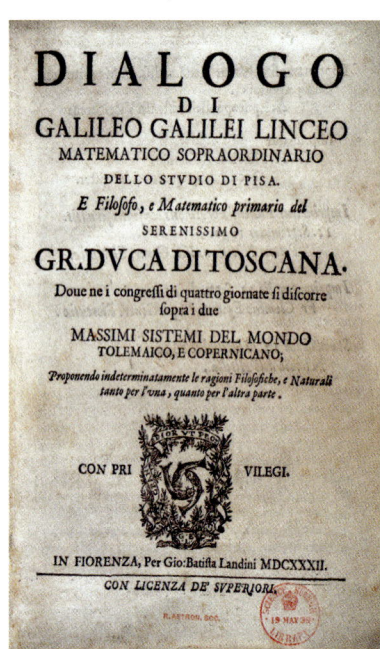

GALILEO'S THEORIES

Galileo was the first person to articulate clearly the principle of relativity. In 1632, he said that the laws of motion and mechanics were the same for any observers who were moving at a constant velocity relative to one another. There was no way to declare that someone was absolutely "at rest" in the universe and another person was "moving," only that they were moving relative to one another. Galileo described this vividly in his *Dialogue Concerning the Two Chief World Systems*. He sought to defend the Copernican thesis that Earth is not a motionless body around which everything else in the universe revolves, but that our planet itself is in motion.

LEFT: *The title page of Galileo's* Dialogo, *which he first published in 1632.*

Maxwell which describe the behavior of electromagnetic waves, such as light. Maxwell found that they had to travel at a certain speed: approximately 186,000 miles (299,000 km) per second. This gave rise to the question of exactly what was it that caused the propagation of these electromagnetic waves. Also, their speed had to be regarded as being relative to something, but to exactly what was a mystery.

Scientists posited that there must indeed be an unseen medium that was propagating these waves, and they called it the "ether." The speed of light waves, they assumed, was a speed relative to this ether. Ether, they considered, played a role for light waves analogous to that which air did for sound waves. That meant that if you were zipping through the ether toward the source of light, the waves would seem to come at you faster.

Scientists in Europe and America therefore launched a "great ether hunt." They assumed that moving through it at speed toward the source of the light should cause us to perceive the light waves traveling by us at greater speed. All sorts of ingenious devices and experiments were devised to detect such differences.

In Cleveland in the late 1880s, Albert Michelson and Edward Morley performed some of the most famous such experiments using a device intended to split a beam of light into two parts, one of which would travel in the direction of the earth's movement around the sun, while another part of the beam went in a perpendicular direction. But no matter what devices they and other scientists built, or whatever assumptions they made about how the ether should behave, they could never actually detect it. Moreover, whichever way round they did their experiments, they always found the speed of light to be precisely the same.

That is where matters stood in early May 1905. Then, one day one of the most serendipitous and momentous leaps forward in the history of physics occurred in the unlikely setting of a conversation between Albert Einstein and Michele Besso.

So what was his insight? It involved, he later said, an analysis of the idea of time itself. Using mental pictures and questioning things other people took for granted, he realized that time was not something that could

BELOW: *A portrait of Galileo by A. S. Zileri after an original by Justus Sustermans. Galileo has been hailed as the father of modern science.*

be defined in absolute terms. In particular, he came to realize that even if one person might perceive two events as happening at precisely the same time, another observer moving at a different speed to the first might not perceive them as simultaneous at all.

Einstein explained this by using a thought experiment, which we can imagine him formulating as he sat in the Patent Office. Picture the scene: he is pondering applications for devices that use signals to synchronize clocks while the trains outside his window rumble under the great Bern clock tower into the station with its rows of identical clocks along the platforms. Here is his thought experiment: suppose lightning bolts strike both ends of a fast-moving train. If a man standing on the platform halfway between the two strikes sees the light from each strike at the exact same moment, he would say that the strikes were simultaneous. But now let us imagine how it looks to a woman at the midpoint inside the train. In the nanosecond it takes the light from the bolts to get to her, she will have moved forward a tiny amount. The light from the front strike will get to her first. She would say they were not simultaneous.

According to Einstein's theory, we cannot say that one of them is "right" and the other is "wrong," because there is no way to declare that one of them is "standing still" and that the other is "moving." Neither of them is standing absolutely motionless with respect to the rest of the universe or has a privileged vantage point. All we can say is that they are moving relative to one another, and that two observers moving relative to one another may have different but equally valid views on what events are simultaneous. His June 1905 paper went on to explain the consequence of this insight. Two events which might seem, just by examining a traditional system of coordinates, to be simultaneous, could no longer be considered so, if the point where one event took place was in motion relative to the position of the other event.

This deceptively simple principle was also revolutionary. It abolished the notion of absolute time. Instead, all systems in motion have different times relative to one another. "This was a change in the very foundation of physics, an unexpected and very radical change that required all the courage of a young and revolutionary

> **"This was a change in the very foundation of physics, an unexpected and very radical change that required all the courage of a young and revolutionary genius."**
>
> —Werner Heisenberg

genius," wrote Werner Heisenberg, whose ideas about quantum uncertainty would similarly turn traditionally accepted notions about physics on their head. The comforting certainty of a time that has an absolute reality and progresses along second by second, unaffected by our observations or perceptions of it had been a bedrock of scientific orthodoxy ever since Newton postulated it 218 years earlier. "Absolute, true, and mathematical time, of itself and from its own nature, flows equably without relation to anything external," he famously wrote in *Book One of the Principia*.

Einstein's genius was that he was willing to challenge what seems to be such an obvious assumption. Then the rebellious patent examiner dismissed two generation's worth of accrued scientific dogma. The light ether so many scientists had sought, was, he stated, an irrelevance.

One consequence of this theory is that time slows down as you move very fast. For a boy trying to catch up with a light beam, time would slow down as he approached the speed of light, which helped to explain why the beam would always race away at

ABOVE: *The world system according to Copernicus from the 17th century.*

BELOW: *An illustration of Einstein's thought experiment on light and its relative spreed from moving objects.*

the constant speed of light relative to him. OK, it is not intuitive, but that is why most of the physics establishment did not grasp it right away. Even after he published his "Miracle Year" papers, Einstein still could not find a job at a university or even one teaching at a high school.

At the beginning of his Miracle Year in 1905, Einstein had written to his friend and fellow member of the Olympia Academy, Conrad Habicht. In September of that year, he sent Habicht another letter, saying that he had come up with a supplemental notion. This, it turned out, would lead to the most famous equation in all of physics. Almost as an afterthought, he added that there must be a direct relationship between the mass of a body and the energy contained within it.

Einstein's expression of this was elegant in the extreme: the energy (L) emitted by a body as radiation causes its mass to decrease by L/V^2. In other words, $L=mV^2$. Einstein employed the term "L" to denote energy until 1912, when he substituted it with "E." Similarly, Einstein at first represented the speed of light by the letter "V" before altering this to "c." Using these new letters, which soon became widely accepted, he now had what would become his best-known equation:

RIGHT: *German physicist Werner Karl Heisenberg. His theories, like Einstein's, would force the scientific community to reexamine ideas they had previously held as sacred.*

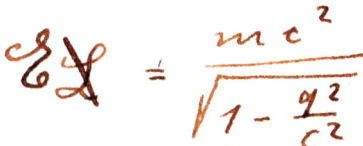

$$e\mathcal{L} = \frac{mc^2}{\sqrt{1 - \frac{q^2}{c^2}}}$$

EINSTEIN'S WALK WITH BESSO

It was a sunny day in Bern, as Einstein would later remember, when he called on Michele Besso, one of his closest friends. Einstein had met this talented but somewhat directionless engineer in Zurich, and had later secured a place for him at the Swiss Patent Office. Einstein had been discussing with Besso his puzzlement about catching up with a light wave, Maxwell's equations, and the mystery of the undetectable ether; at one point he said he was going to give it up. But as they talked about it, Einstein recalled, he suddenly understood the key to the problem. When Besso next saw Einstein the following day, his friend was almost euphoric, announcing that he had completely solved the problem.

LEFT: *Albert Einstein at his desk in the Bern Patent Office in 1905, during his "Miracle Year."*

The Rising Professor

ABOVE: At the same time as Einstein's scientific career was taking off, his personal life was also undergoing significant changes. Five years after this picture of Mileva, Einstein, and the young Hans Albert was taken, Mileva gave birth to a second son.

Einstein's 1905 burst of creativity transformed physics. He had declared that light was a particle as well as a wave, discarded the concept of absolute time, and come up with the most widely recognized formula in physics. But the academic community, while curious, did not rush to offer him a job.

A handful of physicists did take note of Einstein's papers, and one of these turned out to be Max Planck, the person who was—at least up to that moment—the greatest theoretical physicist in the world. Planck was an editorial-board member of the journal that published Einstein, and the paper on relativity had "immediately aroused my lively attention," he later recalled. He gave a lecture on relativity at the University of Berlin and then wrote his own paper building on it.

Einstein was soon corresponding with Planck, who in the summer of 1907 sent his assistant Max Laue to seek him out in Bern. Soon afterwards, Einstein was finally awarded his doctorate, which allowed him to be promoted from a third-class to a second-class examiner at the Patent Office. But he was still not respected enough to be offered a job at a university.

Whereas previously Einstein's isolation from the mainstream of academic physics had been something of an advantage, now it began to act as a drag on his further progress. When he was commissioned to write a major yearbook piece on relativity in 1907, he warned the editor that he might not be aware of all the literature since his work at the Patent Office detained him during the hours the library was open, and thus he could not read all the publications on the subject.

Because he was unable to proceed directly to a professorship, Einstein applied for a position at the University of Bern as a *privatdozent* or tenured professor. He would be required to give some lectures, but, in the absence of an official salary, would collect fees from the students who attended

RIGHT: *Max von Laue, German physicist, winner of the 1914 Nobel Prize for Physics and close friend of Einstein.*

MAX VON LAUE AND HIS MEETING WITH EINSTEIN

When he came to Bern, Laue discovered, much to his surprise, that Einstein was still working at the Patent Office. At first he did not even recognize him. "The young man who came to meet me made so unexpected an impression on me that I did not believe he could possibly be the father of the relativity theory," Laue said, "so I let him pass." Their conversations, as they walked, were lengthy and stimulating. "During the first two hours of our conversation he overthrew the entire mechanics and electrodynamics," Laue remarked. So taken was he with Einstein that they became close friends, and Laue would publish eight papers on the theory of relativity.

them. In applying for this lowly post, Einstein included no fewer than 17 papers which he had already had published, his momentous pieces on relativity and light quanta amongst them. He should also have sent in a thesis known as a habilitation, which had to be unpublished, but, since the authorities allowed this additional requirement to be dropped in the case of applicants with "other outstanding achievements," and as Einstein firmly considered himself to be in this category, he did not bother with it.

He did not get the job. The committee voted not to waive the requirement or to hire Einstein unless he wrote a new thesis. Einstein, ever defiant of authority, refused.

Instead, he tempered his aspirations to be a university professor and began, astonishing though it may seem, to apply for employment as a high school teacher. He explained his desire to join the teaching profession as a means to find some way of continuing his personal scientific work in easier circumstances. He responded to an advertisement for a mathematics and geometry teacher at a high school in Zurich, noting in his application that he was disposed to teach physics as well if necessary. Once again, Einstein included with his application a complete list of his published papers, not omitting that on the special theory of relativity. Faced with competition, from 20 other applicants, Einstein failed miserably even to make it into the final three on the shortlist.

Even a man as proud as Einstein now recognized reality and penned the habilitation thesis necessary to become a *privatdozent*. It was instantly accepted, and so, in February 1908, he finally penetrated the citadel of academe which had so long resisted his entry. Even then, the remuneration was so modest that he could not afford to abandon his position at the Patent Office. His University of Bern lecture course became just one more of his many obligations.

Einstein had still not found complete acceptance in the academic community. Alfred Kleiner, the Zurich physics professor who had been instrumental in helping

Einstein secure his doctorate, and who was considering offering him a professorship at Zurich, was not overly impressed when he attended some of his lectures at Bern.

Anti-Semitism also played a part. Some members of the faculty who were worried by the fact that Einstein was a Jew received the rather telling reassurance from Kleiner that he did not show any of the "unpleasant peculiarities" popularly attached to Jewishness:

"The expressions of our colleague Kleiner, based on several years of personal contact, were all the more valuable for the committee as well as for the faculty as a whole since Herr Dr. Einstein is an Israelite and since precisely to the Israelites among scholars are inscribed (in numerous cases not entirely without cause) all kinds of unpleasant peculiarities of character, such as intrusiveness, impudence, and a shopkeeper's mentality in the perception of their academic position. It should be said, however, that also among the Israelites there exist men who do not exhibit a trace of these disagreeable qualities and that it is not proper, therefore, to disqualify a man only because he happens to be a Jew. Indeed, one occasionally finds people also among non-Jewish scholars who in regard to a commercial perception and utilization of their academic profession develop qualities that are usually considered as specifically Jewish. Therefore, neither the committee nor the faculty as a whole considered it compatible with its dignity to adopt anti-Semitism as a matter of policy."

Thus, Einstein was offered his first professorship, four years after he had revolutionized physics. He accepted it and resigned from the Patent Office. He claimed with some glee to a colleague that he had, in effect, prostituted himself by becoming an academic.

The move to Zurich was extremely welcome to Marić, who remembered their earlier happiness there. She fell pregnant almost immediately, and in July 1910 she bore Einstein his second son, whom they called Eduard (although he was always to be known as Tete.) Einstein was later a somewhat distant father, and particularly towards Eduard, who would suffer from severe mental problems as he grew up, but he was much more indulgent towards his sons in their infancy. "When my mother was busy around the house, father would put aside his work and watch over us for hours, bouncing us on his knee," Hans Albert later recalled. "I remember he would tell us stories—and he often played the violin in an effort to keep us quiet."

It was less than six months after his arrival in Zurich that Einstein received a job offer from the German section of Prague University. It was a significant promotion for Einstein, to a full professorship, but it would cause disruption to his family life. Setting aside such merely personal qualms, Einstein put his career first. He took the job and uprooted his family.

LEFT: *Prague University. Einstein taught theoretical physics here from 1911 to 1912.*

ALFRED KLEINER EVALUATES EINSTEIN

When Kleiner came to attend Einstein's lecture, the younger man did not put on a satisfactory performance. Einstein lamented to a friend that his performance had not been good, the result of his insufficient preparation, but also that being observed had rather irritated him. Kleiner sat impassively, with a faintly concerned look, and once the lecture was over, took Einstein aside to explain that his teaching method was not up to that required for the professorship. Kleiner's criticism probably had some merit. Einstein was never a smooth teacher, and he spoke in a rather disorganized fashion, a trait that, once his celebrity status had become assured, came to be regarded as somewhat endearing. But Einstein asked for another chance, and Kleiner invited him to Zurich for a guest lecture. Einstein told a friend that this time he had been fortunate, since he had lectured uncharacteristically well.

Elsa Einstein

Einstein's climb up the academic ladder was accompanied by invitations to prestigious conferences. As he wandered around basking in his rising renown, Mileva Marić, forced to stay back in Prague, suffered an increasing sense of alienation as she was excluded from just those scientific circles which she had yearned so long to be a part of. "I would like to have been there and listened a little, and seen all these fine people," she wrote him after one of his talks. "It is so long since we saw each other that I wonder if you will recognize me."

Her loneliness in Prague accentuated her natural tendency to depression. Their marriage was already in danger when a meeting Einstein had during a trip to Berlin at Easter 1912 placed it even further in peril. His cousin Elsa Einstein, three years older than him, was living there, and during his visit he made her acquaintance once more.

Elsa was Einstein's cousin on both his parents' sides. As children, they had played together. Since then, Elsa had had a checkered personal life, with a failed marriage that left her (at 36 years old) and her daughters, Margot and Ilse, living in the same apartment block as her parents.

Elsa offered what Einstein now craved. The love he was seeking was not wild romance but uncomplicated support and affection. She and Mileva were very much opposites: Mileva was glamorous, academic and less drawn to the practical; Elsa was none of the

BELOW: *Einstein with Elsa, after she became his wife.*

EINSTEIN'S JOB OFFER

When Planck and Nernst traveled down to make Einstein the offer of twin positions at the Prussian Academy and Berlin University, he said that he needed a little time to consider, although he had almost certainly already decided to accept. The Berlin professors, together with their wives, took themselves off on a short excursion up a nearby mountain. With an impish sense of humor, Einstein told them to look out for him on their return to the funicular railway platform: if they saw him bearing a bunch of white flowers, they would know he had refused the position; but if he had decided positively, they would be red. To their great relief, when they alighted from the train, they were greeted with a red bouquet.

above. Instead, she was a homemaker, with a taste for heavy Teutonic cuisine, and a penchant for chocolate which gave her figure a rather robust look.

When Einstein returned to Prague, Elsa wrote him at his office, not his home, and proposed a way they could send each other letters in secret. Einstein was delighted at her suggestion, going on to admit the fondness he had started to feel for her.

Elsa may not have been a profound intellect, but she had a profound understanding of people. She knew how to make Einstein defensive by teasing him for being under Marić's thumb and calling him "henpecked." As she predicted, Einstein retorted that he was anything but and urged her to dismiss such thoughts. Coyly, he added that he hoped to prove his masculinity to her whenever the occasion might arise.

Nevertheless, he was still trying to please Marić and make their marriage work. So after just a short stint in Prague, he grabbed the opportunity to return to the one place where his marriage might have a chance to survive. In June 1911, the Zurich Polytechnic, where Einstein and Mileva had been so happy as undergraduates, became a full university. As an alumnus, and now one of the world's most renowned theoretical physicists, Einstein was the obvious choice for one of the new professorships available there.

Their move back to Zurich in July 1912 should have been a happy moment. No longer confined to a small apartment, now their resources stretched to six rooms and a glorious view. Evening musical soirées and the company of old friends should have lifted Marić's mood, but instead she became more and more depressed.

Thus, Einstein was receptive when he received a flirtatious letter from Elsa wishing him well on his 34th birthday. In it, she added a request for a picture of him and a recommendation of a good book she could read on relativity. She knew how to flatter: in his reply, Einstein expressed the view that there was nothing worth reading on the subject for the layperson but that, were Elsa ever to visit Zurich, he would take her for a walk (naturally without his wife), and explain to her all about his discoveries in the field. Einstein did not hold back; he suggested that meeting and seeing each other would be better than a mere photograph. He longed to spend time with Elsa, without the crucifying presence of Marić. He enquired if Elsa might be in Berlin over the summer, as he would like to pay a brief visit.

He ended up with an opportunity to do more than merely visit. Two pillars of Berlin's scientific establishment, Max Planck and Walther Nernst, came to see him in July 1913 with an enticing offer. They wanted him to move to Berlin to be a member of the Prussian Academy of Sciences and a professor at the University of Berlin. That evening, having accepted their offer, Einstein wrote in great excitement to Elsa about the huge honor of the new position. He told her that he would transfer permanently to Berlin by the following spring, and that he was already filled with excitement at their being together for longer periods. They began to write more intimate letters to each other. Elsa was concerned that Einstein was overdoing things, and she counseled him to take more exercise, eat better, and rest more. He retorted that he

ABOVE: *The letter in which Einstein offers to explain relativity to his cousin Elsa and asks her to come and visit. (See Translations, page 156.)*

zukommen, wenn ich den ganzen Tag arbeite, ohne mir irgend welche Erholung zu gönnen. Noblesse oblige – mit dieser Berühmtheit ist ein gewisses Elend verbunden.

Wenn Du mir eine grosse Freude machen willst, dann richte es einmal ein, dass Du Dich einige Tage hier aufhältst.

Mit den besten Grüssen, auch an Deine Kinderchen
    Dein
        Albert.

intended on the contrary to smoke more, work more, eat more, and only take exercise if the company was amenable.

More seriously, though, Einstein warned Elsa, that she should not hope for him to leave Marić or divorce her, even after he moved to Berlin. He felt that he and Elsa could be happy together without hurting his wife. Elsa disagreed. She wanted to be Einstein's new wife, not merely his mistress. It would be a bone of contention between them for years, a battle that Elsa would, in the end, win. For the time being, however, Einstein was adamant. He told Elsa that his almost total separation from Marić made the legal nicety of divorce an irrelevance.

The prospect of moving to Berlin depressed Marić deeply. She dreaded the thought of dealing with Einstein's mother, who had never liked her, and his cousin, whom she rightly suspected of being a rival. In a letter to Elsa, Einstein lamented that Marić did nothing but complain about the prospective move and his family in Berlin, though he admitted that she did have something of a point.

Through the spring of 1914, the Einsteins' marriage spiraled downward. The end came in July, when Marić moved with her two boys into the house of a friend. Einstein delivered a brutal ultimatum, revealing both his scientific mindset and emotional coldness, decreeing what she would have to do if she wanted to remain married to him. The terms were harsh, and seemed unfeeling. Marić was to attend to her husband's physical well-being by doing his laundry, cooking for him, and keeping his study neat, while neglecting her own emotional needs, as Einstein forbade her even to sit with him at home and to talk to him unless he requested it. He insisted she leave his presence when asked. Intimacy between them was out of the question.

Marić soon realized that their marriage could not be saved. They met on a Friday to work out a separation (though not yet a divorce) agreement. When the meeting was over, Einstein went to Elsa's apartment. She was on holiday in the Bavarian Alps with her two daughters, and Einstein wrote to her telling her he was now sleeping in her bed. He confided to her that he found it comforting sleeping there, though he chided himself for his sentimentality.

Marić and their two sons left for Switzerland on July 29, 1914. They caught the morning train to Zurich, and in the afternoon Einstein cried like a child. It was a disturbingly personal feeling for a man who prided himself on his detachment from such emotions. He believed himself unaffected by lasting attachments to others, but he was deluding himself. His love for Mileva Marić and his children had been very real, and their departure grieved him deeply.

It was not long before a battle broke out between the estranged spouses over financial support and Mileva's alleged attempts to turn the boys against Einstein. It was not the only conflict which was to affect Einstein profoundly, for that autumn Europe became engulfed in World War I. Its barbarity served as yet another external threat from which Einstein took refuge by submerging himself in his scientific work.

RIGHT: *A special edition of a Berlin newspaper announcing the start of World War I in 1914.*

THE BEGINNING OF WORLD WAR I

The outbreak of war in autumn 1914 left Einstein nursing a sense of profound bewilderment, particularly as he now found himself working in the very country that had unleashed a war of aggression on its neighbors. Always an instinctive pacifist, he found it particularly hard to stomach that many of his scientific colleagues at Berlin University seemed to be enthusiastic proponents of the war. Nernst, who had recruited him to the university, worked on Germany's development of poison gas, while a group of 93 German academics—including Max Planck—and intellectuals signed an "Appeal to the Cultured World" seeking to justify Germany's position and deny that it had been the aggressor. Einstein's reaction to all this was twofold. On the one hand, he joined the "New Fatherland League," an organization dedicated to making an early peace and setting up structures to make a future war impossible. On the other, he threw himself into his scientific work with redoubled dedication.

Extra-Ausgabe

Berliner Lokal-Anzeiger

An jeder Anschlagsäule ist die nächste Geschäftsstelle des „Berliner Lokal-Anzeigers" angegeben.

32. Jahrgang.　　　　　　　　　　Freitag, 31. Juli.　　　　　　　　　　1914.

Kaiser Wilhelm an sein Volk.

Begeisterte Kundgebungen vor dem Schloß. — Der Reichstag wird einberufen. — Die Bekanntmachungen des Oberbefehlshabers in den Marken.

In ernster Stunde.

Eine erlösende Kraft liegt in der Gewißheit. Das Ungewisse ist's immer, das die Herzen im Banne der schwersten Sorge hält. Diesen Satz darf man wohl auf die Stimmung anwenden, die in ernster Stunde jetzt unser deutsches Volk beherrscht. Der stärkste Optimist vermag wohl jetzt kaum noch an der Gewißheit der kommenden welterschütternden Ereignisse zweifeln. Sicherlich sind sie nicht dazu angetan, Freudenstürme auszulösen. Dennoch wird wohl allenthalben in deutschen Landen empfunden werden, daß dieses lange und bange Schwanken zwischen Sorge und Hoffnung fast unerträglich wurde. Haben wir es doch alle neuerdings erlebt, wie dieser Zustand auf unser soziales und wirtschaftliches Leben einen Einfluß ausübte, den Düsteres noch düsterer färbte und alle die Unbilden, die triegerische Verwicklungen uns bringen können, mit Hilfe der erregten Phantasie bis ins Ungeheuerliche vergrößerte.

Nun, da wir — wenn nicht Unerwartetes sich ereignet — an der Schwelle blutiger Entscheidungen stehen, können wir wohl unser Haupt in aufrichtiger Betrübnis darüber, daß die Jourdige, bis an die Grenze des Menschenmöglichen gehende Ruhe und Gemessenheit, mit der unser Kaiser und seine Regierung von uns und von andern das Schlimmste abzuwenden trachteten, anscheinend vergeblich war. Aber mit mannhafter Ruhe und Zuversicht schauen wir dem, was ohne unsere Schuld unabwendbar war, ins grimme Antlitz! Wir sind bereit, die herben Schicksalsschläge zu tragen, die selbst ein siegreicher Kampf dem zu der Erfüllung heiliger Pflichten das Schwert ziehenden Volke schlägt. Wir können es mit gutem Gewissen, wenn noch ein Funken von Glauben an göttliche Gerechtigkeit in des deutschen Volkes Seele lebt. Kein politischer Sophismus, wenn er nicht wider besseres Wissen zu klügeln vermag, vermag die Wahrheit aus der Welt zu schaffen, daß hier ein wahrhaft friedliebendes, in heißer, stiller Kulturarbeit ringendes, wenn auch allezeit wehrhaftes Volk durch Drohung zur Gegendrohung gezwungen wurde. Diese Tatsache wird, des sind wir gewiß, dem Urteilsspruch der Welt, das Weltgericht ist, einstmals zugrundeliegen.

Aber auch ein anderes Fundament hat unsere stolze Zuversicht gegenüber dem uns wohl zur Gewißheit Werdenden. Das ist das unerschütterliche Vertrauen zu unserm Heer und zu unserer Flotte, an deren materieller und moralischer Stärke unsers geliebten Kaisers eiserner Wille und die Opferwilligkeit des gesamten deutschen Volkes mitgearbeitet haben. Denn, den tapferen Söhnen unseres Volks, die — wenn's das nicht nur unabänderlich beschlossen hat — in freudiger Erfüllung unserer übernommenen bundesbrüderlichen Ehrenpflicht und in Verteidigung der Ehre und Sicherheit des Vaterlandes zum Kampf sich bereit machen, wendet sich in der ernsten Schicksalsstunde unser dankbares Herz zu, auf sie sehen wir getrost unsere Hoffnung!

Bängliches Zagen und blasse Furcht lassen wir beiseite. Zusammenschließen wollen wir uns über allen kleinlichen Hader hinweg, in dem Bewußtsein, daß das Schicksal der kommenden Tage der Entscheidung einer uns allen gemeinsamen großen und guten Sache gilt.

* * *

Auf das Rückseite dieser Extraausgabe geben wir die bisher herausgegebenen, für das Publikum wichtigen amtlichen Publikationen über die durch Anordnung des Kriegszustandes eingetretenen besonderen Verhältnisse wieder.

Einberufung des Reichstags.

Für den Fall des Kriegsausbruchs ist die Einberufung des Reichstags auf Dienstag, 4. August 1914, in Aussicht genommen. Die Eröffnung wird im Weißen Saal des Königlichen Schlosses zu Berlin um 1 Uhr nachmittags erfolgen. Die kaiserliche Verordnung wegen der Berufung steht noch aus.

Da der Reichstag bei der Entscheidung über Krieg und Frieden nicht mitzusprechen hat, wird er sich versammeln, um offizielle Mitteilungen über die Entschließung des Kaisers und des Bundesrats entgegenzunehmen und eventuell die Kosten für die Kriegsführung zu bewilligen. Da die Session im Frühjahr geschlossen wurde, muß bei Neubeginn, bevor es in materielle Verhandlungen eintritt, ein neues Präsidium wählen, aber angesichts der Lage der Dinge dürfte sich die Wahl diesmal sehr schnell und glatt vollziehen.

Die Verkündigung des Kriegszustandes in Bayern.

Telegraphische Meldung.

München, 31. Juli.

Nach einer Königlichen Verordnung vom 31. Juli 1914 wird über das Gesamtgebiet des Königreiches der Kriegszustand verhängt. Für die Pfalz wird das Standrecht angeordnet.

Eine Kundgebung der württembergischen Regierung.

Drahtbericht unseres Korrespondenten.

w. Stuttgart, 31. Juli.

Der Staatsanzeiger gibt heute nachmittag 2 Uhr 45 Minuten folgendes Extrablatt aus: „Nachdem Se. Majestät der Kaiser das Reichsgebiet in Kriegszustand erklärt hat, hat Se. Majestät der König das feste Vertrauen aus, daß die Zivilverwaltungen und Gemeindebehörden ihre nur im Interesse des Vaterlandes einzutretende Unterstellung unter die Militärbefehlshaber in einem dem Moment gerecht werdenden gehobenen Pflichtbewußtsein aufnehmen und ihre obliegenden Amtsaufgaben mit besonders freudigem Diensteifer gewissenhaft erfüllen werden."

Auf Allerhöchsten Befehl sind dem Staatsministerium vorstehendes bekannt.

Maßnahmen der Zivilbehörden.

Wie wir hören, hat der Minister des Innern angeordnet, daß alle beurlaubten Beamten im Bereich des Ministeriums des Innern sofort zurückzurufen sind. Demgemäß haben das Polizeipräsidium und die anderen dem Ministerium des Innern unterstellten Behörden-Großberlins ihre Beurlaubten, wo dies nötig war, telegraphisch zurückgefordert. Fast alle der in den Ferien weilenden Beamten sind bereits in den Dienst zurückgekehrt.

Eine Ansprache des Kaisers.

Der Zustrom des Publikums nach den Linden nahm während der Nachmittagsstunden immer mehr zu. Von allen Seiten kamen Tausende und aber Tausende, Männer, Frauen und Kinder aller Stände, und wandten sich nach dem Schloßplatz, der bald nicht mehr zu passieren war, da die Menschenmauer undurchdringlich geworden war. Der Fuhrverkehr hatte schon lange eingestellt werden müssen. Das Schutzmannsaufgebot versuchte vergebens, wenigstens eine schmale Gasse aufrechtzuerhalten. Wohl 50,000 Mann brachten vor dem Schloß fortgesetzt brausende Hochrufe auf Kaiser Wilhelm und das Deutsche Reich aus; dazwischen ertönten patriotische Gesänge, die wieder von Hochrufen abgelöst wurden. Gegen 6¼ Uhr erschien der Kaiser am Fenster, und man konnte die Begeisterung der Menge keine Grenzen mehr. Hüte und Mützen und Tücher wurden geschwenkt, es wurde in die Hände geklatscht.

Der Kaiser spricht.

Alle schauten mit Spannung zum Kaiser empor, der nun mit Ehrn und mit weithinschallender Stimme eine Ansprache hielt, deren wörtlicher Wortlaut folgender ist:

„Eine schwere Stunde ist heute über Deutschland hereingebrochen. Neider überall zwingen uns zu gerechter Verteidigung. Man drückt uns das Schwert in die Hand. Ich hoffe, daß, wenn es nicht in letzter Stunde meinen Bemühungen gelingt, die Gegner zum Einsehen zu bringen und den Frieden zu erhalten, wir mit Gottes Hilfe so führen werden, daß wir es mit Ehren wieder in die Scheide stecken können. Enorme Opfer an Gut und Blut würde ein Krieg von uns erfordern. Den Gegnern aber würden wir zeigen, was es heißt, Deutschland zu reizen. Und nun empfehle ich euch Gott. Jetzt geht in die Kirche, kniet nieder vor Gott und bittet ihn um Hilfe für unser braves Heer!"

Dreimal nahme diese kurze Ansprache des Kaisers durch jubelnde Zurufe des Publikums unterbrochen. Nachdem der Kaiser ausgeredet hatte, betont dann wieder das Sturm des Schloßzimmers, begleitet von den brausenden Hochrufen der begeisterten Menge, die noch, nachdem der Kaiser sich schon zur Ansicht umgekleidet hatte, anhielt. Überall wurde in erregten Gruppen die Rede des Monarchen kommentiert, und überall fühlte man, daß sie dem Volke zu Herzen gedrungen war.

Unter den Linden.

Die vaterländische Begeisterung des Publikums verdichtete sich zur Nachmittage zu außerordentlichen Kundgebungen. Auch die Bekanntmachungen des Oberbefehlshabers in den Marken, die den Kriegszustand proklamierten, angeschlagen wurden, kam es zu Ausbrüchen des Patriotismus. Unter den Linden wehten sich die Massen in fast beängstigender Weise. Beim Schloß entfaltete sich um 4 Uhr nachmittags in eigenartiger Schauspiel durch die

Proklamierung des Kriegszustandes.

Eine Abteilung des Kaiser-Alexander-Garde-Grenadier-Regiments marschierte unter dem Kommando des Bataillonsadjutanten Leutnants von Biebahn die Linden entlang zum Zeughaus. Dort wurde „Halt!" kommandiert. Der Leutnant verlas die Bekanntmachung des Oberkommandierenden in den Marken und proklamierte

unter Trommelwirbel den Kriegszustand.

Die eilig hinzuströmende stark angewachsene Menge brach in donnernde Hurrarufe aus. Wo sich ein Mitglied der Königlichen Familie zeigte, wurde es mit patriotischen, herzlichen Zurufen begrüßt. Sobald das kaiserliche Automobil-Signal ertönte, stürzten die Massen nach der Fahrstraße. Tücher wehten. Die Hurrarufe pflanzten sich brausend fort und schwollen an, sobald die Kraftwagen dem Schloße kamen. Auch der Reichskanzler, der um 4 Uhr ins Schloß zum Vortrag zurückkehrte, wurde der Menge erkannt und herzlich begrüßt. Die französische und russische Botschaft wurden zu zu zeinerlei Kundgebungen, wie überhaupt Ziel und Zweck der ersten Tausende, die nachmittags Unter den Linden versammelt waren, in der Hauptsache die herzliche Begrüßung des Kaisers und seiner Familie bildete. Gegenstand warmherziger Ovationen waren allenthalben auch der Kronprinz und die Kronprinzessin.

Die Bekanntmachungen an den Anschlagsäulen.

Um die Anschlagsäulen in Berlin drängte sich bereits 4 Uhr eine vielköpfige Menschenmenge. Fünf Bekanntmachungen des Generaloberst von Kessel, des Oberbefehlshabers in den Marken, macht erklärt sind, daß der Kriegszustand in Berlin eingetreten sei, und worin die näheren Bestimmungen über die Durchführung enthalten sind, wurden gelesen und machten auf alle erzählsftliche Eindruck. Von den Omnibussen und den Straßenbahnen sprangen die Leute herunter und eilten zu den Säulen. Einzelne lasen die Bekanntmachung laut vor, andere wieder erläuterten einzelne Bestimmungen. In der Markgrafenstraße brachte ein ordensgeschmückter Invalide den begeistert aufgenommenen Hoch auf den Kaiser aus. Auch in den Arbeiterquartieren war dies nicht anders. Die Spannung hat gewichen, einer Begeisterung machte noch geltend, wie die wirkungsvollen und nachhaltigen kann durch die Stadt wechselnden. Alle Soldaten und Reservisten geben ihren Begeisterung Ausdruck. Der alle Furor teutonicus war erwacht! Die große Stunde fand Berlins Bevölkerung groß.

Die Sicherheit der Spareinlagen.

Angesichts der an einzelnen Orten auftretenden Besorgnisse der Bevölkerung wegen ihrer Spareinlagen in den städtischen Sparkassen hat der Minister des Innern unter dem 30. v. M. allgemein bekanntgegeben, daß nicht zu irgendeiner Beunruhigung besteht. Für jede öffentliche Sparkasse haftet ihre Stadt oder ihr Kreis oder der sonstige Kommunalverband, der sie errichtet hat, mit seiner gesamten Steuerkraft. Die Gelder der öffentlichen Sparkassen sind für den Falleines Krieges als kriegssichere im absoluten Sinne anzusehen. Jedem Zugriff des eigenen Staates sowohl wie des Feindes entzogen, die öffentlichen Sparkassen bieten daher den Einlegern die denkbar größte Sicherheit, und es kann den Sparern nur empfohlen werden, in dem Falle auch ihres Krieges alles verfügbare Geld dort nieder-

General Relativity

ABOVE: *Einstein in 1914, a year before he announced his findings about the general theory of relativity at the University of Göttingen.*

Einstein's first step toward a general theory of relativity came in November 1907, when he was writing an article explaining his special theory of relativity. Two limitations of the theory bothered him. First, it said that no physical interaction can propagate faster than the speed of light; that conflicted with Newton's theory of gravity, which conceived of gravity as a force that acted instantly between distant objects. Second, it applied only to the special (hence the name) case involving constant-velocity motion. If you were accelerating or turning or slamming on the brakes, things seemed to behave differently.

Einstein, being a good scientist, didn't like it when things applied only to special cases. So he embarked on interwoven quests to generalize relativity so that it applied to accelerated motion and to come up with a new theory of gravity.

Einstein later recalled that a sudden insight had come to him while seated in the Bern Patent Office; that if a person were to fall freely, he would feel himself weightless. It was a startling realization and was the first step in this quest. Later, he would consider this as the happiest thought he had ever had.

Einstein extended his thought experiments to other cases involving the effects of gravity and acceleration. He imagined a man who was standing in an enclosed chamber in a gravitational field, such as on the Earth's surface. What would he feel? He would feel (it is easy to imagine) his feet being pressed to the floor. If he took something out of his pocket and dropped it, the object would fall to the floor at an accelerated rate. Then Einstein imagined the man being in the same chamber deep in outer space where there was no gravity. But the chamber is being accelerated upward. What would he feel? The same thing! His feet would be pressed to the floor. If he drops an object, he will watch it fall to the floor at an accelerated rate.

Einstein called this the principle of equivalence. The effects of being in a gravitational field are equivalent to those of being accelerated upward. And as with the woman on the train and the man on the platform who perceive differently the timing of two lightning strikes, neither way of perceiving things is right or wrong. He termed this the equivalence principle and it led him to a profound insight; that both gravity and acceleration are produced by the same underlying phenomenon.

In 1911, he described one consequence of the equivalence principle: gravity should curve a light beam. This is apparent from a development of Einstein's thought experiment about the chamber. If a pinhole in the chamber allows a beam of light to enter the chamber from one wall whilst it is being accelerated upward, the beam will arrive at the opposite wall at a lower point, since the chamber itself will have moved a little upward. Plotting the

> **"Every boy in...Göttingen understands more about four-dimensional geometry than Einstein. Yet... Einstein did the work and not the mathematicians."**
> —David Hilbert

trajectory of the light-beam in this situation would reveal a curved shape, precisely because of this upward acceleration. Since Einstein's equivalence principle dictated that acceleration and a gravitational field will produce the same effects, then light must similarly display a curved trajectory when it passes through a gravitational field.

These insights led Einstein to a totally new concept of gravity. It was not, as Newton had said, some mysterious attraction between two objects at a distance. Instead, gravity is a phenomenon in which objects cause the fabric of space to curve, and this curving of space determines the movement of objects.

Years later, when his younger son Eduard asked why he was so famous, Einstein replied using a simple image to describe his great insight that gravity was the curving of space-time. He likened it to the motion of a blind beetle crawling over a branch. The beetle does not know that the branch is curved. His discovery of gravity's curvative effect was simply his chance observance of what the beetle did not notice.

Describing the curvature of space-time required mathematical tools that went beyond the elegant geometry of Euclid that Einstein had admired as a schoolboy. Unfortunately, as he showed at the Zurich Polytechnic, non-Euclidean geometry was not a strong suit for Einstein. Fortunately, his old friend Marcel Grossmann stepped in to assist him.

When Einstein moved to Berlin in the spring of 1914, and after his wife and sons left him to return to Zurich, he rented an apartment that was near to Elsa. He lived a solitary, almost monastic life, eating and sleeping at odd times, dedicating his entire being to what he called the *Entwurf* or "outline," a solution to the general relativity problem. But as its name suggests, it was incomplete. The more that Einstein examined the mathematics of this version of the theory, the more it seemed to fall short.

Einstein decided to describe the status of his quest at a weeklong series of lectures, which began in late June 1915. He chose the

DAVID HILBERT (1862–1943)

Born in Königsberg (modern Kaliningrad, Russia), Hilbert became a professor at Göttingen University in 1895, where he was to spend the rest of his career. He was initially noted as a mathematician, working across a wide range of subjects. In 1900, he set forward at a Paris Congress a list of 23 unsolved problems, which in many ways established the agenda for the next century of mathematical work. He sought to put mathematics on a firm axiomatic footing, a goal which set him in opposition to Kurt Gödel's "intuitionist" school. His venture into theoretical physics in 1915, prompted by his contact with Einstein, created severe tensions with the physicist, who feared his work was being usurped. They later became reconciled, and Hilbert helped Einstein to be elected to the Göttingen Academy. Although a political conservative by instinct, Hilbert also shared Einstein's views on Germany's role in World War I and was one of the scientists who signed the "Appeal to the Cultured World."

University of Göttingen as the venue, because of its leading reputation in the mathematics of theoretical physics. In a letter to another physicist, Einstein was effusive about the place. He was delighted that everyone there seemed to grasp the details of what he was saying. The one in the audience he was particularly interested in was David Hilbert, whose overeagerness to have the new relativity theory explained to him presaged trouble to come. Hilbert was so taken with Einstein and the special relativity problem that he decided to see if he could solve it himself. A frantic scramble followed to figure out the mathematical field equations to describe the theory. When Einstein heard of this competition, he was distressed. A frenzied month ensued for Einstein, as he desperately sought to outpace his rival, giving four lectures to the Prussian Academy on the latest results of his struggles with the equations. In November 1915, however, Einstein finally emerged triumphant, with a definitive revision of the Newtonian universe.

The Prussian Academy met weekly in formal sessions in the State Library's grand hall, at which the members would listen to their peers giving forth on the latest results of their researches. Einstein's first address to it, on general relativity, was delivered on November 4. In it he explained how his scientific labors for the past four years had been directed toward extending his theory of relativity to a general one, even where motion in the system observed was not uniform. He said he was confident of the

THE CURVATURE OF SPACE

Einstein again came up with a vivid way to visualize the curvature of space. Imagine rolling a bowling ball onto a two-dimensional fabric, such as the surface of a trampoline. It curves the fabric. Now roll some billiard balls on it. What do they do? They soon start rolling toward the bowling ball. Why? Not because, as Newton tried to explain it, the bowling ball has some mysterious attraction at a distance. It is because the bowling ball has curved the fabric and the rolling billiard balls follow that curve. We can easily visualize this in two dimensions, like on a trampoline. Einstein could visualize an object curving all three dimensions of space. In fact, what he actually visualized was how an object curves all four dimensions of the intertwined fabric of space and time, known as space-time.

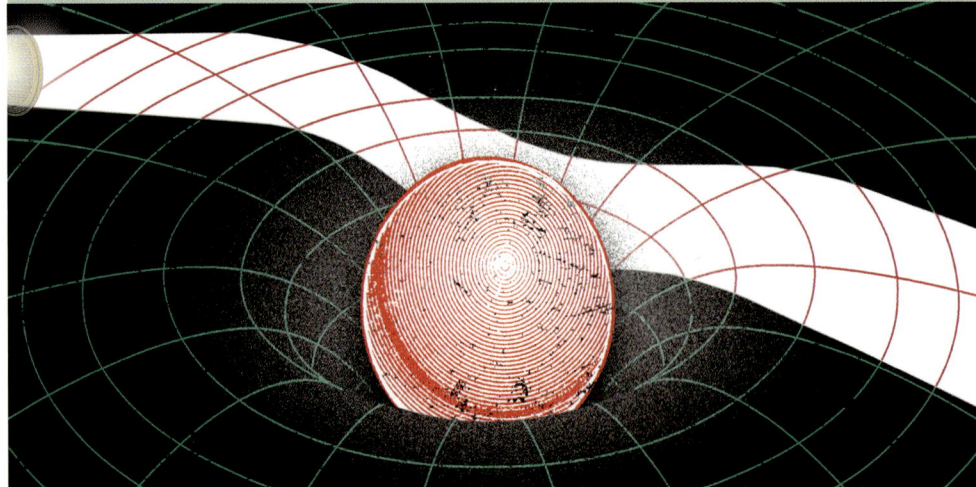

ABOVE: *While developing his general theory of relativity Einstein still took an interest in the work of his friends, as this 1911 letter from Fritz Haber indicates. (See Translations, pages 156–157.)*

scientific basis for his theory, but he admitted that he had not quite got the mathematical equations nailed down.

Einstein was in the throes of the one of the most concentrated frenzies of scientific creativity in history. He was working intensely. In the midst of this, he was also still dealing with the crisis within his family. Letters arrived from his estranged wife that pressed the issue of his financial obligations and discussed the guidelines for contact with his sons. On the very day he turned in his first paper, November 4, he wrote an anguished—and painfully poignant—letter to his estranged son Hans Albert, promising that he would try to visit for a month each year, and to be a real, loving father. Hans Albert, he added, could learn a lot from him that no one else could teach the boy. He explained that in the last few days he had put the finishing touches to a paper of which he was most proud, and that he looked forward to explaining it to his son when he was older. He apologized to Hans Albert for his distraction, saying that he was often so caught up in his scientific work that he totally forgot about lunch.

Einstein's relations with Hilbert now took a turn for the worse. Einstein was very worried that Hilbert would use his work, overtake him, and manage to reach a general theory of relativity first. Einstein wrote to his rival, enclosing a copy of his November 4 lecture. Defensively, he enquired whether Hilbert thought his new solution was an effective one. In the paper he delivered the following week, Einstein refined the mathematical tensors he was employing. He appeared to be close to the final solution but was frustrated by his slowness in completing its final stages. To protect himself, he also sent a copy of that paper to Hilbert.

Hilbert's response was hardly designed to allay Einstein's fears. He professed himself committed to finding a "solution to your great problem." He had planned to refine his own work further, "But since you are so interested, I would like to lay out my theory in very complete detail this coming Tuesday." He extended an invitation to Einstein to visit Göttingen and hear Hilbert delivering the results of his own work at a public lecture to be held November 16. Nonchalantly, he even enclosed the times of the afternoon trains from Berlin and added: "My wife and I would be very pleased if you stayed with us." Tantalizingly, Hilbert finished his letter with a provocative postscript: "As far as I understand your new paper, the solution given by you is entirely different from mine."

PAUL DIRAC (1902–1984)

Born in Bristol of a Swiss father, Dirac, who was destined to become one of Britain's greatest theoretical physicists, studied electrical engineering at Bristol University, graduating in 1921. At Cambridge, where he studied for his doctorate, his interests turned to the new field of quantum mechanics. He pioneered the area of quantum electrodynamics, and developed equations which predicted the existence of the positron. His work was recognized by his appointment in 1932 as Lucasian Professor of Mathematics at Cambridge and by the Nobel Prize for Physics in 1933. From such a distinguished physicist, it was high praise indeed that he should have said of Einstein's general theory of relativity that it was "probably the greatest scientific discovery ever made."

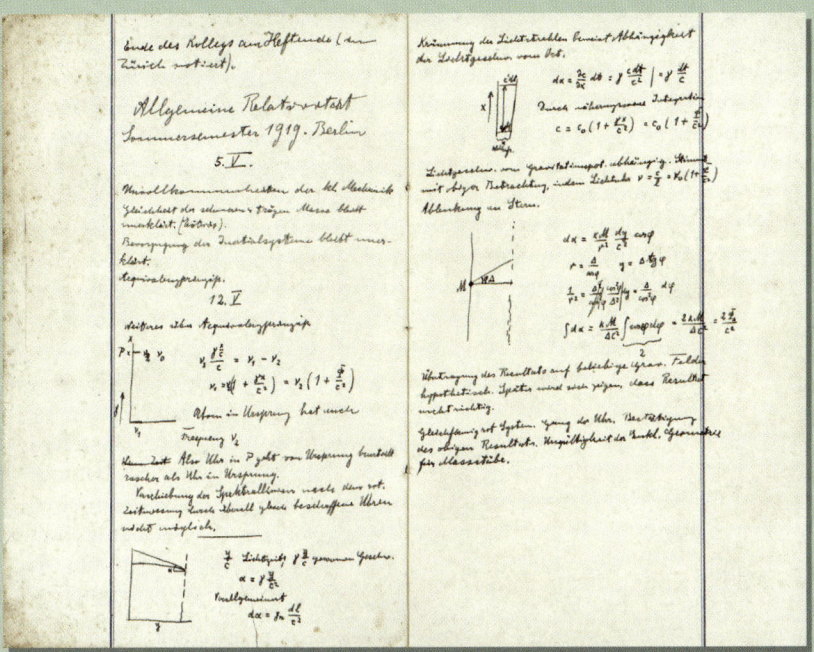

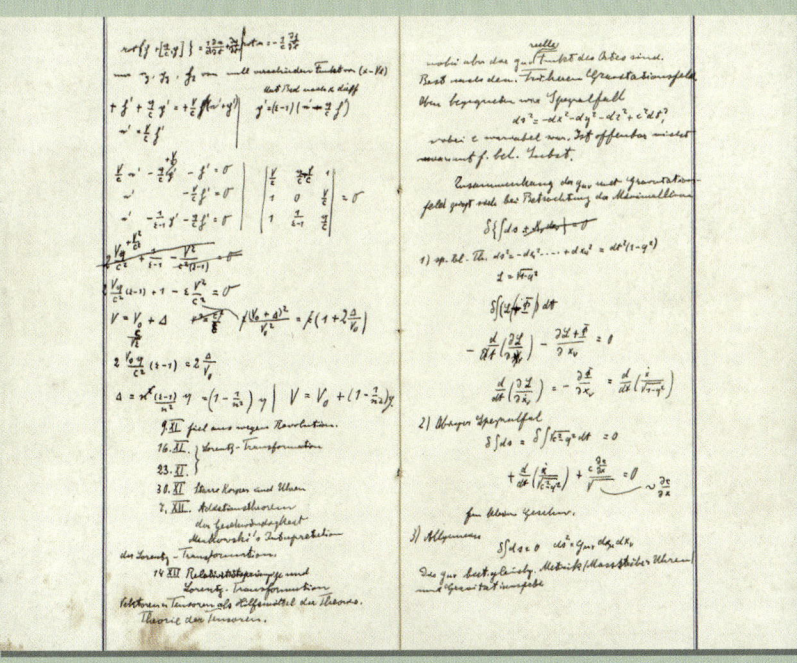

ABOVE: *Einstein's handwritten notes for his lectures on relativity at the Universities of Zurich and Berlin in 1918 and 1919. (See Translations, page 156.)*

Einstein wrote a series of letters on November 15 that give an insight into quite how busy his life had become, so stressful that he had begun to suffer stomach pains. He wrote once more to Hans Albert, saying that he intended to come and visit him in Switzerland at Christmastime. He also sent a letter to Mileva, thanking her for not seeking to damage his relationship with his sons. And he reported to a friend that he had reformulated his gravitational theory, realizing that his earlier proofs were defective. He added that he would be visiting Switzerland at the end of the year.

Einstein also wrote to Hilbert. His letter was full of anxiety, refusing an invitation to visit Göttingen the next day. Einstein wrote that exhaustion and stomach pains prevented him from visiting, but that he would be grateful if Hilbert could send him some proofs of his work, as he was intrigued by the hints contained in his letters.

Einstein's mood was lifted by a discovery he made that week. Even though he knew there was work to be done to perfect his equations, he decided to turn them to something that had been bothering physicists: an unexplained tiny aberration in the orbit of the planet Mercury. The result was a triumph, and one he revealed to the Prussian academicians at his third lecture: his new theory of relativity accounted for the tiny shift in Mercury's orbit. He was so thrilled he had heart palpitations. To another physicist he exulted that the detected movement of Mercury vindicated him and that he was profoundly grateful to the astronomers whose insistence on measuring minuscule differences he had previously mocked.

Einstein also used the lecture on November 18 to announce an update to a calculation he had made eight years earlier, when he began working on the general relativity theory. He had predicted that the sun's gravitational field would bend light by 0.85 arc-seconds. With his new equations, Einstein revised this to an amount twice as great. To test the new figure he would have to wait for an eclipse more than three years in the future.

The very same morning, Einstein received a copy of the paper which Hilbert had invited him to Göttingen to hear. It seemed disturbingly close to his own research. Einstein responded brusquely, clearly asserting that he had reached the same conclusions first and independently. He wrote to Hilbert baldly stating that the results in his rival's papers agreed with what he had already presented to the Academy. Einstein went on to say he was about to present another paper in which he would explain the perihelion motion of Mercury in terms of general relativity. It was something, he could not resist adding, that only his own gravitational theory had managed to achieve.

Hilbert's response was generous and did not claim priority for his own conclusions: "Cordial congratulations on conquering Mercury's motion," he wrote. "If I could calculate as rapidly as you, in my equations the electron would have to capitulate and the hydrogen atom would have to produce its note of apology about why it does not radiate." Yet Hilbert's apparent modesty must have seemed like dissembling, when

MARCEL GROSSMANN HELPS EINSTEIN AGAIN

When Einstein found himself foxed by the non-Euclidean geometry he needed to define the gravitational field, his old friend Marcel Grossmann, whose mathematics notes at the Polytechnic had so assisted him in countering his own tendency to skip the classes, once again came to the rescue. Whereas Einstein had scraped by with 4.25 out of 6 in the two courses on geometry, Grossmann had achieved a maximum 6 and, most usefully for Einstein's present difficulties, had submitted his dissertation on non-Euclidean geometry, and subsequently published seven papers on the subject. Einstein implored Grossmann to help him, or else he felt he would lose his mind. From 1912 to 1915, Einstein, with Grossmann's help, wrestled with mathematical tools known as "metric tensors" in an effort to find the field equations that would mathematically convey his insight that gravity could be defined as the curvature of space-time.

just the next day he submitted a paper which contained his own general relativity equations to a science journal in Göttingen. The title he chose had more than a hint of braggadocio: "The Foundations of Physics."

No one knows how attentively Einstein read the paper Hilbert sent to him, or whether he derived any insights from it in his preparation for his momentous fourth lecture to the Prussian Academy. More likely, it was the work that he had done the week before on Mercury and the curvature of light that led him to his own conclusions. So it was that by November 25, 1915, the date of the final lecture, which he entitled "The Field Equations of Gravity," he had produced new equations which set the seal on the general theory of relativity.

RIGHT: *A copy of On the Special and General Theory of Relativity by Einstein. This copy was the first one to leave the printer in 1917.*

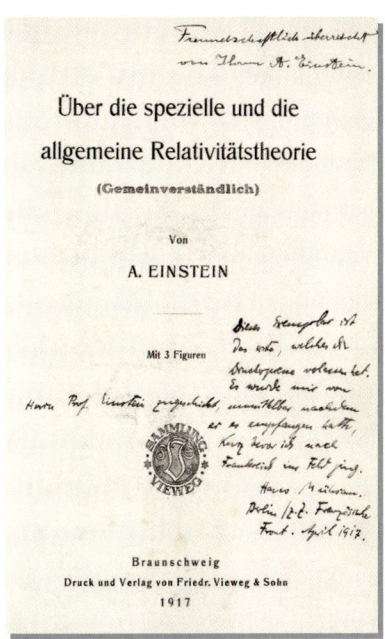

To the layperson it would not seem nearly as dramatic as the elegant simplicity of $E=mc^2$. Yet the shorthand notation of tensors, which transforms abstruse mathematical notions into inoffensive subscripts, condensed Einstein's final equations into a compact enough form even to be printed on T-shirts, although the sort more often sported by physics undergraduates than the average follower of fashion. In one, almost visually digestible form, it is written as:

$$R_{\mu\nu} - \tfrac{1}{2} g_{\mu\nu} R = 8\pi T_{\mu\nu}$$

The left side of the equation describes how the geometry of space-time is warped and curved by matter. Its right side deals with how matter moves within the gravitational field. The two sides of the equation interact to demonstrate the curvature of space-time by objects, and how, conversely, the movement of those objects is affected by that curvature. In the concise explanation of the physicist John Wheeler: "Matter tells space-time how to curve, and curved space tells matter how to move."

Einstein was worried that Hilbert would be accorded some of the credit. But Hilbert graciously declared that it was Einstein's theory. "Every boy in the streets of Göttingen understands more about four-dimensional geometry than Einstein," he said. "Yet, in spite of that, Einstein did the work and not the mathematicians."

At just 36, Einstein had come up with one of the most revolutionary revisions of our ideas about the universe that had ever been achieved. The Newtonian universe had been one in which time was an absolute concept, wholly independent of, and unaffected by, observation, and in which space, too, had a monolithic and tangible quality. In Newton's worldview, gravity was seen as something exerted on each other by bodies, in a mysterious and not entirely explicable way, across space. Einstein's special theory of relativity changed all this. He showed how time and space were interrelated. With the general theory, he took this even further, to show how space-time had a particular set of dynamics, which was affected by objects, and in turn had an effect on the motion of objects within it.

Max Born, another titan of 20th-century physics, referred to Einstein's work as "the greatest feat of human thinking about nature."

Einstein was worn out but jubilant. Despite the worries of a broken marriage and a continent blighted by war, he was probably at his happiest. In a letter to Besso he exulted that his most audacious dreams had been made a reality and, when he signed the letter, he wrote that he was happy but exhausted.

General Relativity 69

The Home Front

Einstein's great strength as a scientist was his nonconformity. He refused to accept authority or convention. That was true not only in his science. It was also evident in his political outlook and in his personal life.

When war broke out in Europe in 1914, it inflamed the patriotic pride of the Prussians. Einstein, on the other hand, proclaimed his pacifism and became a leader of the international war resisters' movement. Indeed, there are few more stark examples of nonconformity than deciding to become a pacifist and a war resister in Berlin in late 1914.

Einstein believed that scientists had a duty to oppose the war. It was irrational, he felt. In fact, he believed that all forms of nationalism were irrational. He argued that it was the responsibility of the scientific community to foster internationalism, but he noted that, sadly, some scientists had failed to do this. That is why he was dismayed that his three closest scientific colleagues at the University of Berlin—Fritz Haber, Walther Nernst, and Max Planck—fell in lockstep to Germany's military mentality.

All three signed a petition supporting Germany's cause in the war. Einstein responded by signing a pacifist declaration that could garner only two other signatures. He also became an early member of the pacifist "New Fatherland League," which worked for the achievement of an early peace and the setting up of a federal system in Europe that would avoid any such future conflicts. Einstein's two sons were still living with his estranged wife in Zurich, and the war made the separation more difficult. Hans Albert, who was just turning 11, wrote him two heart-wrenching letters begging him to come visit him and Eduard for Easter in 1915. "I just think: At Easter you're going to be here and we'll have a Papa again." In his next postcard, he said that his younger brother Eduard told him about having a dream "that Papa was here." He told his father how well he was doing in mathematics at school. "Mama assigns me problems; we have a little booklet; I could do the same with you as well."

The difficulty of wartime travel made it impossible for him to visit at Easter, and so Einstein did not get to Switzerland until early September. Marić still hoped that she and Einstein might be reconciled, and she asked him to stay with her and the boys at her apartment, but Einstein had no desire to be with her. Instead, he stayed on his own, spent most of his time with his friends, and saw his sons only twice during the three weeks he was in Switzerland. He wrote to Elsa, putting the blame for this on Mileva, who, he claimed, was afraid that the boys would come to depend on him too much.

Later that autumn, during the tense weeks of November 1915 when he was racing with Hilbert to finalize the field equations of general relativity, Einstein's son Hans Albert was telling a family friend of his desire to spend some time

RIGHT: *The Krupp cannon foundry in Essen in 1914. This factory provided the German army with heavy artillery during World War I.*

FRITZ HABER
(1868–1934)

Fritz Haber was born Jewish but tried, to Einstein's dismay, to assimilate into German society. He converted to Christianity, underwent baptism, and affected the dress of a thoroughly Prussian gentleman, even down to pince-nez glasses. A renowned chemist, he would win the Nobel Prize in 1918 for finding a way to synthesize ammonia from nitrogen, but this was turned by the German military to the production of explosives on an industrial scale. He also participated in the development of chlorine gas, whose deadly clouds would deal an agonizingly painful death to thousands of soldiers, their lungs and throats choked with the lethal, burning substance. Haber was even present in person at Ypres in April 1915 when the age of chemical weapons was inaugurated, with the resultant death of 5,000 French and Belgian soldiers. His conversion to Christianity would avail him nothing when the Nazis came to power, and he was forced to flee Germany in 1933, dying of a heart attack the next year in Switzerland.

WALTHER NERNST
(1864–1941)

Another of Einstein's colleagues at Berlin who disappointed him by his attitude to the war, Walther Nernst graduated from the University of Wurzburg in 1867 with a thesis on electromotive forces. He became professor of chemistry at Berlin in 1905, and his groundbreaking work on thermochemistry yielded the Third Law of Thermodynamics and was recognized by a Nobel Prize in 1920. At the age of 50, Nernst decided to volunteer to be a driver on the German front. He practiced his marching style and military salutes in an impromptu parade in front of his wife. As an academic rival of Haber, he did work on tear gas and other chemicals which could be used as a nonlethal way of clearing soldiers from trenches. The army preferred Haber's much more lethal substances, and, once that decision had been made, Nernst joined in the development of poison gases.

with his father at Christmas. Yet at the same time he was sending his father an unpleasant letter asking him to refrain from coming to Switzerland at all. Unsurprisingly, the boy had very ambivalent feelings towards his father, who had, after all, virtually abandoned his family.

Einstein, so patient and persistent when teasing out scientific problems, mirrored this trait with an equal and opposite refusal to spend time on difficult personal problems. He told Hans Albert that he would not come to Switzerland after all. He also balked when he found out that he was supposed to pay for a set of skis that Marić had bought Hans Albert for 70 francs as a Christmas present. "Mama bought them for me on condition that you also contribute," he wrote. "I consider them a Christmas present." A vexed Einstein responded that he would send Hans Albert cash as a present, but he added that they really could not afford expenditure on such extravagances.

Marić tried to smooth things over with a conciliatory letter, and so Einstein decided, in the end, that he should visit Zurich. It did not come to pass. In the wake of his triumph in conquering general relativity, he was exhausted. The war, too, had made cross-border trips a difficult enterprise, and so on December 23, when he was due to set out for Switzerland, Einstein wrote to Hans Albert, saying that the visit would not, after all, be possible. He justified his change of heart because of the wartime uncertainties of a border crossing and the great strain of his work which left him feeling exhausted.

Einstein spent a solitary Christmas Day in his Berlin apartment. Some drawings that his son had sent him caught his attention, and he wrote a postcard to Hans Albert saying how much he had enjoyed looking at them. He promised to come and visit at Easter, and he asked his son to keep studying the piano, so that his father could join him in a duet on the violin.

BELOW: *German stormtroopers appearing out of a cloud of poison gas they had laid while attacking British trenches. Haber played a vital role in the development of chorline gas.*

Divorce and Remarriage

ABOVE: *Lucerne as it would have looked when Einstein and Hans Albert visited it in 1916.*

Just as he told Hans Albert he would, Einstein arrived in Zurich early in April 1916 for a three-week Easter break. He took up residence in a hotel near the railway station. The visit went well at first. His sons seemed genuinely pleased to see him, and Einstein took Hans Albert on a 10-day hike in a mountain resort near Lake Lucerne. However, they were trapped in their hotel by a storm. Einstein wrote to Elsa relating how the pair had become snowed in but were enjoying themselves greatly. Einstein found Hans Albert's questioning and his curiosity particularly gratifying. Yet the enchantment of their enforced closeness soon palled and Einstein cut the trip short.

Hans Albert begged him, at the very least, to make a brief courtesy call on Marić, who was still legally his wife. But Einstein was adamant in his refusal to see her. He and his son argued about it one morning when the boy, still just 12, came to visit the Zurich physics institute to watch an experiment. Hans Albert refused to return in the afternoon for the end of the experiment until his father agreed to see Mileva. Einstein, however, would not be persuaded.

Einstein had refrained from seeking a divorce from Marić. He had no desire to get married again. He found the nature of his relationship with Elsa, with its lack of longer-term commitments, most congenial. Elsa, on the other hand, held fast to her resolve to become Einstein's wife. She kept pushing. So Einstein, during that spring of 1916, began to beg Marić to grant him a divorce, so that they could get on with the rest of their lives apart.

The demand for a divorce and her husband's unwillingness to see her during his Zurich visit provoked a sudden downturn in Marić's physical and emotional state. She had a number of coronary scares, began to be afflicted by extreme anxiety, and was advised by her doctors that she needed a prolonged period of bed rest. The two boys moved in with friends.

Einstein hated emotional turmoil. Whenever confronted with a personal crisis, he tended to retreat into science to avoid confrontation. So he decided to give up asking Marić for a divorce, at least for the time being. That seemed to aid her recovery. He told a mutual friend how he had resigned himself to leaving Marić well alone, would cease to bother her about the divorce proceedings, and now felt more able to devote himself to his scientific work.

It was then Einstein's turn to fall ill, after which he moved in to Elsa's apartment building, where she could more easily look after him. There was one catch: Elsa still wanted to get married, and the new living arrangements gave her more leverage. So it was inevitable that Einstein would raise with Marić the issue of a divorce again. In early 1918, he wrote to her again stating that he wished to put his personal affairs in order, and that he was therefore determined to pursue the matter of a divorce.

He raised his offer of how much money he would pay to support her each month. Then he added an astonishing new incentive. It had been 13 years since his breakthrough papers, and the Nobel committee had found various excuses not to honor him. But he was convinced that one of these days he would win the prize, and so he made a proposal designed to turn Marić's head. If Einstein should win the Nobel Prize, he would give the entirety of the prize money to her.

That was a huge offer. It was in excess of 35 times what Marić had previously received from Einstein each year. There was also a sense of belated justice in the offer. Marić had helped Einstein when he was writing his 1905 papers, and she was receiving something in return.

Her first response was anger. "Exactly two years ago, such letters pushed me over the brink into misery," she replied. "Why do you torment me so endlessly? I really don't deserve this from you." A cooler reassessment of Einstein's offer brought a calmer response. She was sickly, depressed, and worried about money. There was little likelihood of winning her husband back. And when she consulted with her friends in the scientific community, it seemed like a good bet that he would win the Nobel sometime soon. So here was a deal that might put an end to her struggles with her husband and financial insecurity. She decided to accept it.

As they worked out the details and filed the necessary papers, both of them expressed some relief that matters were headed toward a resolution. Her health got better, the children moved back in with her, and the correspondence between Einstein and Marić lost the sharp edge of bitterness that had infected it. Einstein's divorce proceedings came to a climax in December 1918, just after the World War ended. Einstein had to give a deposition admitting that he had, indeed, been committing adultery. So he appeared before a court magistrate in Berlin and declared that he had been living with Elsa Einstein, his cousin.

As soon as he could—on June 2, 1919, before, in fact, it was technically permitted by the divorce decree—Einstein married Elsa. Surprisingly, that seemed to calm everyone

EINSTEIN FALLS ILL

Although it was Marić who had previously suffered the worse health, in 1917 it was Einstein who became unwell. His bad eating habits, emotional turmoil, and intense focus on his science as he paced around his lonely apartment caused him to come down with stomach pains. It was hard during the war to get food, and Einstein was not one who would stand in line for his rations. In the early summer of 1917, Elsa came to the rescue. She moved him into an apartment in her building, where she could feed and care for him. She had both the resources and resourcefulness—and the desire—to find the comfort foods he liked, as well as cigars. Einstein, for his part, was very content to be looked after.

ABOVE: *A Berlin soup kitchen in World War I. Einstein's refusal to visit such places or to even claim his own rations led to him falling ill.*

Divorce and Remarriage 75

ABOVE: *The letter Einstein wrote to Mileva offering her the money from the Nobel Prize (which he believed he would win) in return for a divorce. (See Translations, page 157.)*

ELSA THE WIFE

For Einstein, his second marriage was the mirror image of his first. It was not romantic. He and Elsa did not share a bedroom, choosing to sleep in different parts of their huge Berlin flat. There was none of the emotional or intellectual passion that marked his early relationship with Marić. Whereas Marić brooded about being denied her chance to shine as a scientist, Elsa joked that she was just the opposite. Understanding relativity, she said, "is not necessary for my happiness." But Elsa, unlike her new husband, had practical talents, which is what made her complementary as a companion. She could manage a household, handle logistics, and translate for him in French and English. "I am not talented in any direction except perhaps as wife and mother," she said. "My interest in mathematics is mainly in the household bills."

down, including Marić. Einstein visited his sons in Zurich, and he stayed at his ex-wife's flat while she was away. Elsa was not exactly happy about this, but Einstein mollified her by saying that her former rival would be away most of the time, and that there would be no chance of any unfortunate encounters with her bête noire. Einstein and Hans Albert made up for lost time by doing many of the things his absence had made impossible, such as going sailing, enjoying music together, and even constructing a model airplane. Einstein wrote to Elsa how Hans Albert's determination and precision gave him great pleasure. His son's virtuosity on the piano especially delighted Einstein.

Elsa would have found her new role in Einstein's life a difficult one, although not without its rewards. Understanding relativity was challenging, but so too was playing with good humor the role of wife and mother to Einstein, who required both. And managing their increasingly complex lives—letting him bask in celebrity while shielding him from its demands—was a tricky feat that she performed with good sense and warmth. She displayed an unaffected manner and self-aware wit, and in doing so she thus helped make sure that her husband retained those traits as well.

There was a symbiosis to their relationship, which meant that the desires of both partners were satisfied. Elsa wanted above all to serve him and shield him. He was eager to be served and protected. She rather enjoyed his famous stature and was entirely open about this, a contrast to her husband's diffidence on the subject. She relished the enhanced social status that his increasing recognition gave the couple.

During her husband's periodic bouts of intense scientific activity, Elsa would cook Einstein's favorite dishes—lentil soup and sausages—and call him away from his study. He chose, though, to eat his meals alone, so engrossed in his thoughts that the act of eating was a mere mechanical one. Elsa packed for him before his many lecture trips, and ensured he had enough money to meet his needs. Publicly, she was very defensive of him, referring to him as "the Professor" or just as "Einstein." The bubble she created around her husband allowed him to unravel the secrets of the cosmos undistracted by the world around him. From this, Elsa derived the most profound satisfaction.

DAILY SKETCH, TUESDAY, NOVEMBER 12, 1918.

CRUSHING ARMISTICE TERMS: GERMAN NAVY RESIST

DAILY SKETCH.

No. 3,019. Telephones: {London—Holborn 6512. Manchester—City 6501.} LONDON, TUESDAY, NOVEMBER 12, 1918. [Registered as a newspaper.] ONE PENN

THE KING AND HIS PEOPLE HAIL VICTORY DA

Victory Day! A glorious day in British annals, hailed with a joy and pride and thankfulness transcending any within living memory. At Buckingham Palace there wa remarkable demonstration, recalling the memorable moment on the eve of the war pictured on Page 5, as the King, in Admiral's uniform, with the Queen and Princess M beside him, stepped on to the balcony and saluted the cheering multitude. "With you," said His Majesty, "I rejoice and thank God for the victories which the Allied Ar have won and brought hostilities to an end and peace within sight." And the cheers that rolled through London were echoed in the furthermost posts of Empire.

The Eclipse

LEFT: *The edition of the British newspaper the* Daily Sketch *announcing the end of World War I. The Armistice made it possible for Eddington to test Einstein's theory during the 1919 eclipse.*

When it became clear that they were losing World War I, many Germans blamed the situation on the pacifists, the internationalists, and the Jews. Einstein fell into three out of three of these categories. His work was labeled "Jewish physics," and it was criticized for being too theoretical and not rooted in observable reality.

The best way to refute that criticism was to conduct a real-world test to see if his theories were correct. For general relativity, there was a dramatic test possible, one which might help distract a war-scarred world from its wounds. Moreover, its basic principle was so simple that the general public would be able to understand it: gravity would bend a light beam. In particular, Einstein had calculated the amount that the sun's strong gravitational field would be seen to cause the light from a distant star to curve as it passed through it.

Einstein issued a call for scientists to test his theory in a popular book he wrote in 1916. He had calculated that the effect of gravity should curve the light from stars by 1.7 arc-seconds, and issued a challenge to astronomers to prove or disprove his assertion by observation. There was one big complication: in order to make

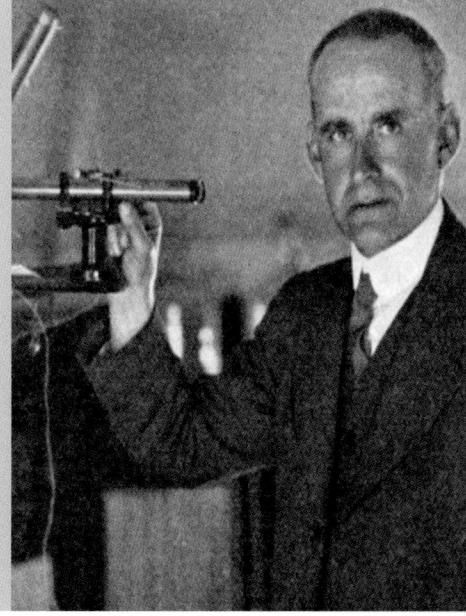

SIR ARTHUR EDDINGTON (1882–1944)

Born into a traditional Quaker family in the north of England, Arthur Eddington would become one of Britain's leading theoretical physicists and astrophysicists and a leading early supporter of Einstein's theories. A Cambridge graduate, in 1906 Eddington became chief assistant to Frank Dyson, the Astronomer Royal, at the Greenwich Observatory. In 1913, he returned to Cambridge as professor of astronomy and director of the observatory there, in which capacity he planned and accompanied the 1919 eclipse expedition. He spent much of the rest of his career seeking a theory which would unite relativity, quantum mechanics, and gravitation into one overarching system. Although he failed in this, and his "fundamental theory" was only published posthumously in 1946, his endeavors inspired generations of scientists in their search for a unified theory.

the observation, there needed to be a total eclipse, so that the sun's light would not blot out the stars and they could be photographed. Such total eclipses, however, do occur relatively frequently in locations that made such an experiment practicable.

Even before the World War had ended, Sir Arthur Eddington, the director of the Cambridge Observatory, decided to take up the challenge. A Quaker English pacifist who needed a way to justify his refusal to join the military, Eddington set out to prove the theory of a Jewish German pacifist. It was his intention to show that science triumphed over politics.

The next great opportunity would be during an eclipse that was due to occur in May 1919. The sun was due to pass through the Hyades, a dense cluster of stars close to the middle of the constellation Taurus. Observing the eclipse would not be exactly easy, since the path of its maximum visibility would be a band in the Atlantic close to the equator from the Brazilian coast to that of east Africa. When Eddington was planning his mission in 1918, the area was patrolled by German U-boats, whose captains had more interest in the arc of their torpedoes' trajectories than in that of light passing through a gravitational field.

Luckily for Eddington, the war had ended by the time his two teams sailed from Liverpool in February 1919. One group headed to the town of Sobral in northern Brazil, while the other, which included Eddington, went to the tiny island of Principe, a Portuguese colony just off the Atlantic coast of Africa. The photographic plates from both places had to be sent back to England on a ship, then carefully developed, examined, and compared to each other. The whole process dragged on until September, while Europe's scientists could scarcely conceal their impatience to know the results.

Einstein pretended to be sanguine as he awaited the results in Berlin. But even he could not hide his impatience. In a letter to a friend in Holland, Einstein casually enquired whether he had heard anything about the observation of the solar eclipse. He finally

LEFT: *A diagram that appeared in the* Illustrated London News *on November 22, 1919, illustrating Einstein's theory based on photographs of the eclipse.*

ABOVE: *One of a set of photographs taken by Eddington that confirmed Einstein's general relativity theory.*

LEFT: *At the same time as Einstein forced a rethinking in the scientific community, artists such as Pablo Picasso were changing people's perceptions through works of art.*

LEFT: *Hendrik Lorentz's telegram informing Einstein that his theory of deflection of light by the sun had been confirmed. This proved Einstein's principle of relativity. (See Translations page 157.)*

ABOVE, LEFT: *A picture from the early 1920s with Einstein, Paul Ehrenfest, and Willem de Sitter at the back and Eddington and Lorentz in front.*

ABOVE, RIGHT: *Igor Stravinsky, Russian-American composer and a contemporary of Einstein's. Stravinsky has been hailed as a musical revolutionary because of his technical and creative innovations.*

got a preliminary report in September, when Hendrik Lorentz sent a telegram to Einstein summarizing what a colleague, who had spoken to Eddington, had passed on to him. Early indications showed that Einstein had been correct.

Ilse Schneider, a graduate student who was with Einstein just after Lorentz's news reached him, described his reaction: "He suddenly interrupted the discussion" and reached over to a window ledge where the telegram was sitting, and casually announced that its contents might be of some interest to her.

Schneider was excited, but Einstein remained calm, maintaining that he had always believed his theory would be vindicated. Schneider asked what he would have done if the experiments had failed to prove his theory. Einstein replied firmly that he would have been sorry for the dear Lord, because the theory was correct.

At the official announcement of the results in November, although there was wide agreement that the theory was historic, most people also considered the whole thing rather baffling. One member of the audience came up to Eddington, who had declared

Einstein's theories to be correct, and said that it was widely held that there were only three scientists in the whole world who actually understood general relativity. Eddington was considered to be among this trio of sages.

For a while Eddington stayed silent. When the man pressed him further, saying "Don't be so modest, Eddington," the reticent Quaker responded: "On the contrary. I'm just wondering who the third might be."

Baffling or not, the theory of relativity captured the imagination of a world thirsting for demonstrations of the positive side of the human yearning for excellence. *The Times* of London announced it with the following:

REVOLUTION IN SCIENCE
New Theory of the Universe
NEWTONIAN IDEAS
OVERTHROWN

"The scientific concept of the fabric of the Universe must be changed," The Times declared. The theory of relativity, so clamorously confirmed by the eclipse measurements, would "require a new philosophy of the Universe, a philosophy that will sweep away nearly all that has hitherto been accepted."

It took two days for the *New York Times* to run the story. It gave the assignment to its golfing correspondent Henry Crouch, as the paper had no scientific writer based in London. The story that he came up with bore one of the all-time classic headlines of newspaper history:

> Lights All Askew in the Heavens
> Men of Science More or Less Agog Over Results of Eclipse Observations
> **EINSTEIN THEORY TRIUMPHS**
> Stars Not Where They Seemed or Were Calculated to Be
> But Nobody Need Worry

There are defining moments in the history of ideas, when a fundamental shift in mankind's outlook occurs. The Enlightenment, when art, philosophy, and politics underwent a profound transformation, saw one such shift. The ideas of the Enlightenment, however, had been molded by the scientific outlook of Isaac Newton and his mechanical universe. His was a cosmos shaped by unshakeable certainties and universal laws, and these notions had been transported into the milieux of politics and philosophy to create an outlook that was profoundly attached to the primacy of cause and effect, of order, and even of duty.

Now, three centuries later, a new way of looking at the universe—relativity—erupted into the political and philosophical consciousness. No longer were space and time seen to be absolute notions. Certainties appeared to be undermined. This loss of faith in the absolute seemed to some to smack of heresy, and even of atheism. As the historian Paul Johnson noted in *Modern Times*, his history of the 20th century, Einstein's theory of relativity "formed a knife to help cut society adrift from its traditional moorings." The result of this shattering of long-held beliefs and debunking of precious truths gave rise to the movement known as modernism. It was not just in science that the strictures of classical thinking were broken. A wave of new thinkers discarded the old ways in almost all fields of human endeavor, including not only Einstein, but Pablo Picasso, Henri Matisse, Igor Stravinsky, Arnold Schoenberg, James Joyce, T. S. Eliot, Marcel Proust, Sergei Diaghilev, Sigmund Freud, Ludwig Wittgenstein, and dozens of others.

This was a misreading of Einstein. His theory of relativity should not, in fact, be confused with relativism, especially with moral relativism. But this was precisely how he came to be viewed in the popular imagination. The modernist trend would all soon lead to some troubling developments and reactions, none more unsettling than in the Germany of the 1920s.

ABOVE: *An oil painting of Marcel Proust. In 1919 Proust published a further volume of his remarkable work* **Remembrance of Things Past.** *In these books Proust examined the themes of time, space, and memory.*

ANNOUNCEMENT AT BURLINGTON HOUSE

The official announcement of the results of the eclipse expeditions came on November 6, 1919. The members of the Royal Society, the most august and venerable of Britain's scientific institutions, gathered in the grand hall of Burlington House for the historic event. "After a careful study of the plates, I am prepared to say that there can be no doubt that they confirm Einstein's prediction," Sir Frank Dyson, the Astronomer Royal, announced. The president of the Royal Society, J. J. Thomson, set the tone for the reaction. "It is the greatest discovery in connection with gravitation since Newton enunciated his principles," he declared. Einstein, still back in Berlin, chose to celebrate by investing in a new violin.

BELOW: *Burlington House in 1925, six years after the official announcement of the eclipse experiment results was made there.*

Einstein in America

Einstein's growing fame as a scientist and his incipient Zionism came together in spring 1921 for a unique event, a scientific tour of the United States, unequaled in the annals of science for the frenzy with which he was received. It was a level of adulation that would have satisfied many latter-day rock stars. No one had ever seen a scientist accorded such celebrity status, and his appearance as an avid promoter of humanist values and as a near hero for Jews, only fanned the fires of his fame.

It all began with a telegram from Chaim Weizmann, president of the World Zionist Organization, which was relayed to Einstein

LEFT: *Kurt Blumenfeld, German author, politician, and Zionist.*

CHAIM WEIZMANN (1874–1952)

Chaim Weizmann was born in Tsarist Russia in the village of Motol (now in Belarus). He settled in Switzerland, where he graduated in chemistry and lectured from 1901 to 1903 at Zurich University. There he became involved in the Zionist movement and began to lobby for the foundation of a Hebrew University. He moved to England to lecture in Manchester and became a member of the General Zionist Council in 1905. After World War I, he was instrumental in the drafting of the Balfour Declaration, which promised a Jewish homeland in Palestine. By the time Einstein accompanied him on their tour of the United States, Weizmann had become president of the World Zionist Organization. He served as first president of the State of Israel from 1949 until his death in 1952.

ABOVE: An anti-Semitic propaganda poster for the 1924 Reichstag election on behalf of the Völkischer Block (national block).

by Kurt Blumenfeld, the German Zionist leader. In it, Weizmann suggested that Einstein come with him to the United States, where he was going on a fundraising trip to help Jews wanting to settle in Palestine and to assist in the foundation of the Hebrew University in Jerusalem. Einstein was not initially keen, claiming that he was no orator and that the idea of exploiting his celebrity was hardly a worthy one.

Blumenfeld responded by reading out Weizmann's telegram once more. "He is the president of our organization," Blumenfeld said, "and if you take your conversion to Zionism seriously, then I have the right to ask you, in Dr. Weizmann's name, to go with him to the United States."

Einstein replied that he was convinced. His eminence linked, like it or not, to the Zionist cause and he would accept the invitation. The decision represented a turning point for Einstein. Until that moment, his life had been almost wholly dedicated to science, even when this had been at the cost of his personal and family life. Yet the longer he lived in Germany, the more he became connected to his identity as a Jew. Anti-Semitism stoked his instincts as a rebellious outsider. Instead of trying to assimilate or hide his Jewishness, Einstein became more assertive about his links to the culture and community of what he called his tribal kinsmen. In correspondence with a friend he noted that he was doing whatever he possibly could for those of his fellow Jews who were being treated so shamefully almost everywhere.

The Einsteins left for America in March 1921. During the crossing, Einstein tried to explain relativity to Weizmann. When reporters in New York asked him whether he understood the theory, Weizmann replied: "During the crossing, Einstein explained his theory to me every day, and by the time we arrived I was fully convinced that he really understands it."

A reporter then turned to Einstein and asked for a single-sentence summary of his theory. In response Einstein quipped that he had spent the whole of his life trying

> **"[I]f you take your conversion to Zionism seriously, then I have the right to ask you, in Dr. Weizmann's name, to go with him to the United States."** —Kurt Blumenfeld to Einstein

to condense it into a single book, and now some journalist wanted a single sentence. When pressed to come up with something, he produced an engagingly simple overview: his work, he explained, dealt with space and time in terms of physics. Its end result was a gravitational theory.

He was then asked questions about his opponents, particularly those in Germany, who had been criticizing his theories. Einstein countered that the only physicists who discounted his theory of relativity were motivated by political concerns. No one who actually understood the subject would hold such an opinion.

And what, the reporters wondered, might those political motives be. The short and stark reply was that the political ideas behind them were anti-Semitic.

The welcome celebrations included a fife-and-drum band from the Jewish Legion and a motorcade that took Einstein, in an open convertible, through cheering throngs in Jewish neighborhoods of the Lower East Side. Einstein and Weizmann had an official welcome a few days later at City Hall, where 10,000 excited spectators gathered in the park to hear the speeches. "As Dr Einstein left," the New York Evening Post reported, "he was lifted onto the shoulders of his colleagues and into the automobile, which passed in triumphal procession through a mass of waving banners and a roar of cheering voices."

Einstein drew packed houses wherever he appeared in Manhattan, despite the fact that he spoke in German about complex physics or stood to one side smiling silently, as Weizmann sought to extract money from the audience for the Jewish settlements in Palestine. "Every seat in the Metropolitan

ABOVE: *Einstein stepping off the SS Rotterdam in April 1921 in New York, during his first trip to America.*

RIGHT: *From left: Chaim Weizmann, New York Mayor John F. Hylan, and Albert Einstein watch a parade in New York.*

THE SENATE DEBATES GENERAL RELATIVITY

During Einstein's visit to the United States, the Senate decided that it would debate the theory of relativity. Unsurprisingly for such a lay assembly, most of the members concluded that the theory was incomprehensible, which it certainly was to them. The most fervent opponents included Pennsylvania Republican Boies Penrose, who was once heard to utter "public office is the last refuge of a scoundrel," and Mississippi Democrat John Sharp Williams, who, when he retired a year later, remarked "I'd rather be a dog and bay at the moon than stay in the Senate another six years." Even Einstein's supporters in Washington were not wholly unequivocal. Congressman J. J. Kindred of New York, who proposed that an account of the relativity theory be placed in the Congressional Record, confessed that "I have been earnestly busy with the theory for three weeks and am beginning to see some light."

LEFT: *Senator Boies Penrose, one of the senators who debated general relativity.*

Opera House, from the pit to the last row under the roof, was filled, and hundreds stood," reported the *New York Times* one day. The tumult in New York lasted more than three weeks.

Einstein then visited Washington, where he met President Warren G. Harding. He also attended a dinner at the National Academy of Sciences, where he sat through a near-endless series of after-dinner lectures. As one professor from North Carolina droned on about his research on hookworms, Einstein turned to face the Dutch diplomat beside him and declared that his other neighbor's conversational skills had just revealed to him a whole new theory of eternity.

In Princeton, Einstein delivered a series of scientific lectures and received an honorary degree "for voyaging through strange seas of thought." Einstein made no compromises with his audience; he scribbled furiously on the blackboard, setting down no less than 125 complex equations while he spoke in German. One student present in the audience summarized the atmosphere to a reporter: "I sat in the balcony, but he talked right over my head anyway."

After one of the lectures, a party was thrown in Einstein's honor. There he uttered one of his most memorable comments, which gave a glimpse into his amused inner confidence. In high excitement, one of those present told Einstein of a set of experiments that had just been conducted which seemed to show that the ether really existed, and that Einstein's theory that the speed of light was constant had been proved wrong. Einstein just would not believe it. So convinced was he that he was right that he coolly replied that while God might be subtle, he was assuredly not malicious. His words were later carved on the mantlepiece of the fireplace near where Einstein had been standing.

Harvard had invited Einstein to visit, but it did not ask him to give lectures. Many saw this as a slight snub that reflected the influence at Harvard of people such as Louis Brandeis and Felix Frankfurter, proudly assimilated Jews who did not readily embrace the more assertive form of Zionism represented by Einstein and Weizmann. Einstein tended to be amused and somewhat disdainful of Jews who tried too hard to assimilate. In a letter he wrote from Harvard to a friend back in Germany, Einstein remarked on the Brandeis attitude, saying that trying too hard to keep non-Jews on their side was a peculiarly Jewish failing.

The Boston area was an appropriate place to make a point about the methods and purposes of education, and Einstein happened to get a chance to do that when he was confronted with a popular game dubbed the "Edison test." Thomas Edison, the famed inventor, was a practical man who believed that the teaching in American colleges was too theoretical. Instead, he felt, they should concentrate on teaching facts. He was 74 when Einstein visited America, and he was getting crankier. He had come up with a test which he set for prospective job applicants, slightly modified according to the post they were applying for.

Around 150 questions long, it included such questions as: How is leather tanned? What country consumes the most tea? What was Gutenberg's type made of?

It was almost inevitable that Einstein would be subjected to what the *New York Times* called "the ever-present Edison questionnaire controversy." At one appearance, a reporter stood up and asked him, from the Edison test, "What is the speed of sound?" Einstein of course understood the propagation of sound waves very well. But he was forced to confess that he did not have such facts on the tip of his tongue, as he could so easily look them up in books. He then enlarged his reply with a more general attack on Edison's educational views. Education, Einstein declared, was more about training students how to think than overloading their minds with rote learning of facts.

Einstein spent most of his time in Boston going to appearances, rallies, and dinners with Weizmann to drum up contributions for their Zionist cause. The *Boston Herald* reported on one fundraising event at a synagogue in Roxbury:

EINSTEIN AND WARREN G. HARDING

When Einstein went to the White House, President Warren G. Harding was asked whether he understood relativity. With commendable candor, Harding smiled and confessed that he was completely baffled by it. The *Washington Post* satirized his reply with a cartoon that portrayed the president poring over a paper entitled "The Theory of Relativity," while Einstein was shown equally foxed by one called "The Theory of Normalcy," which just happened to be what Harding called his philosophy of government. Despite Harding's popularity at the time, his presidency was dogged by a series of administrative and corruption scandals, and, not long after his death from a heart attack in 1923, the Great Depression and the rise of Fascism in Europe would sweep away any sense of the "normalcy" which he had promised.

LEFT: *President Warren G. Harding, who had only taken office just over a month before Einstein arrived in America.*

"The response was electrifying. Young girl ushers worked their way with difficulty through the crowded aisles, carrying long boxes. Bills of various denominations were rained into these receptacles. A prominent [woman] cried out ecstatically that she had eight sons who had been in the army and wanted to make some donation in proportion to their sacrifices. She held up her watch, a valuable imported timepiece, and slipped the rings from her hands. Others followed her example, and soon baskets and boxes filled with diamonds and other precious ornaments."

With the warped notion of celebrity that dominates our culture these days, it is astonishing to look back on the descriptions of the massive parades that greeted Einstein on each stop of his grand tour of America. In Hartford, Connecticut, Einstein's motorcade was made up of more than 100 vehicles, ahead of which marched a band, a group of war veterans, and flag bearers waving American and Zionist standards. More than 15,000 spectators gathered to watch the procession. "North Main Street was jammed by crowds that struggled to get close to shake hands," the newspaper reported. "The crowds cheered wildly as Dr. Weizmann and Professor Einstein stood up in the car to receive flowers." It was not something we could imagine these days for a theoretical physicist.

Einstein's trip to America and his new association with the Zionist cause was part of a significant trend that was transforming the Jewish sense of identity in Europe. In an interview given on his final day in the United States, Einstein said that German Jews had only within the last generation come to see themselves as members of the Jewish people. Before this, they had simply viewed themselves as belonging to a religious community. This feeling was destroyed by the rapid rise of anti-Semitism. For Einstein, there was still something to be gained from the crisis engulfing the Jewish community. He had always found the tendency amongst Jews to adapt and conform rather repugnant, and if the troubles facing some Jews forced others to confront their Jewishness, then something positive, at least, had come out of the present predicament.

RIGHT: *During his tour of America, Einstein also visited President Warren G. Harding at the White House.*

The Nobel Prize

The eclipse observations had made Einstein world famous and confirmed that he had upended Newton's universe. Yet even when he made his triumphal tour of America, he had not yet won the Nobel Prize. For him, it was somewhat inconvenient; he had promised to give the money from the Prize to his former wife Mileva Marić, in return for her agreeing to a divorce, and instead they found themselves fighting over money and child support.

Year after year, Einstein had been nominated for the prize, only to be bypassed for a shifting set of reasons and prejudices. He was proposed for the first time in 1910 by the chemistry laureate Wilhelm Ostwald, who cited his relativity theory. He emphasized that the theory involved fundamental physics, but Einstein's opponents argued that relativity was more of a philosophical theory than a scientific discovery. It was a criticism that was shared by members of the prize committee in Sweden. They were mindful of the mandate in Alfred Nobel's will that the prize should go to "the most important discovery or invention," and the committee felt that relativity theory was neither of those things. More experimental verification of relativity was needed, the committee decided, before it could be worthy of the prize.

If that had been the only objection, then the issues should have been settled by the conclusive analysis of the eclipse observation and their announcement in November 1919. But by then, politics had intervened. The arguments against giving Einstein the prize now had something of a hint of cultural bias and personal animosity about them. The carnival atmosphere which had accompanied Einstein on his trip to the United States, and his status as science's greatest celebrity since Benjamin Franklin received a similar reception in the streets of Paris in 1776, had not been received well by his critics, who accused him of a self-promotion that made him unworthy of a Nobel laureate.

The prize committee's chairman in 1920 compiled an internal dossier justifying their refusal to give Einstein the prize. It gives a glimpse into the political and personal biases

LEFT: *Allvar Gullstrand in 1910, the year before he won the Nobel Prize over Einstein.*

CHARLES-EDOUARD GUILLAUME

Charles-Edouard Guillaume was Einstein's scientific and personal opposite. Educated in Switzerland, he entered the International Bureau of Weights and Measures in 1883 and became its director in 1915, a post he retained until his retirement in 1936. While studying nickel-iron alloys and their coefficient of expansion, he discovered the alloys invar and elinvar, both of which possess very low coefficients. These found use in the construction of precision measuring instruments and for making measuring rods that would guarantee standard measures. His work, on an incomparably more modest scale to Einstein's, was judged more worthy of the Nobel Prize in 1920 than the revolutionary theory of general relativity.

that were hurting Einstein. The seven-page report cited the discredited work of Ernst Gehrcke, a blatantly anti-Semitic enemy of Einstein who led a rally against him in Berlin in the summer of 1920. Another leading anti-Semitic critic, Philipp Lenard, was also waging a crusade against Einstein behind the scenes. The 1920 Prize in the end went to a Zurich Polytechnic graduate, but not the one that history would have expected. It was awarded to Charles-Edouard Guillaume for his work on metal alloys.

By 1921, support was growing, both among the public and within the scientific community, for Einstein to win the Nobel Prize. The number of official nominations Einstein received reached 14, considerably more than any of the competition. "Einstein stands above his contemporaries even as Newton did," wrote Sir Arthur Eddington.

Once again, this sentiment did not extend to the experimentalists who dominated the Prize committee. The 1921 internal report was written by Allvar Gullstrand, who

OSEEN'S SOLUTION

Because the whole question of a prize recognizing Einstein's relativity work was so enmired in controversy, Oseen decided that it would be better to focus on Einstein's even more revolutionary work involving light quanta. So he proposed that the committee give the Nobel Prize to Einstein for "the discovery of the law of the photoelectric effect." Oseen chose his words with great care. The short phrase did more than merely bypass the controversy over relativity. Despite the way it has been described by some historians, Einstein was not in fact nominated for his theory of light quanta, even though this was the theory which underpinned his 1905 paper on the subject. As phrased by Oseen, the nomination was for his discovery of a law, a formulation which bypassed the objections that the prize could not be awarded for theories.

LEFT: Max Born, seated, and from left behind him: William Osler, Niels Bohr, James Franck, and Oscar Klein. The photograph was taken at the Goettingen Bohr Festival in 1922 when Niels Bohr delivered seven lectures about the theory of atomic structure.

ABOVE: *The Nobel Medal. Each medal features the portrait of Alfred Nobel and the image on the reverse changes according to the discipline.*

was ophthalmology professor at Uppsala University. Gullstrand had received the Prize in 1911 for medicine, but his understanding of mathematics and physics was sadly wanting, and he was, moreover, intent on undermining Einstein's claim to the prize. Gullstrand's report claimed not only that the curving of light was not an adequate test of Einstein's theories, but that the results of his experiments were not valid, and that even had they been, the observations could still be explained using Newtonian mechanics.

Although Gullstrand's assault on Einstein's ideas was somewhat crude, as even many members of the Swedish Academy realized, it was put in such a way that it was difficult to oppose. The Academy equivocated, and refused to issue a verdict, something that Einstein might have judged even more damning than awarding the prize to someone else. The Academy's considered judgment was that the 1921 prize be awarded to no one, and held over until the next year.

By this point the situation was more embarrassing to the Swedish Academy and the Nobel Prize than it was to Einstein. Fortunately, in 1922, a theoretical physicist from the University of Uppsala, Carl Wilhelm Oseen, joined the committee. He came up with a way to break the odd impasse, by proposing that Einstein be awarded the prize for his work on light quanta.

Oseen proposed that the committee give Einstein the 1921 prize that had been left on the shelf a year earlier. This would permit the Academy to give Niels Bohr the 1922 prize, on the grounds that his atomic model was based on the laws governing the photoelectric effect. By this clever presentation, the anti-theorist instincts of the experimentalists who were part of the old guard at the Academy were overcome, and the two greatest theoretical physicists of the time became Nobel laureates. In September 1922, the Academy voted to confirm the awards, and so Einstein and Bohr received the 1921 and 1922 Nobel prizes respectively.

Einstein was alerted in advance that he was going to be recognized at the 1922 ceremonies, which were scheduled for December, but by then he was merrily dismissive of the whole process and had decided to proceed with a planned trip to Japan. It was not until the following July that Einstein gave his official acceptance speech, at a Swedish science conference. Although the prize had been given for his work on the photoelectric effect, Einstein chose to give his speech on relativity and, a new interest for him, on the vital importance of developing a unified field theory that would reconcile general relativity, electromagnetic theory, and quantum mechanics.

The Nobel Prize money that year amounted to more than 10 times his annual professorial salary. It allowed Einstein to fulfill, finally, the deal he had made in his divorce agreement with Marić in early 1918. She eventually bought three small apartment buildings in Zurich with the money.

> **"Einstein stands above his contemporaries even as Newton did."**
> —Sir Arthur Eddington

Quantum Mechanics

As the Nobel Prize showed, Einstein had been a pioneer of quantum theory—the idea that light and all forms of radiant energy were composed of discrete particles. But during the mid-1920s, quantum theory produced a radical new system of mechanics that caused Einstein increasing discomfort.

As developed by Werner Heisenberg, Niels Bohr, and others, quantum mechanics held that the dual nature of light—both wave and particle—meant that there was an inherent uncertainty at the subatomic level, that observing a phenomenon played a role in determining its reality, and that there was an element of probability or chance in the way that photons were emitted and electrons behaved. The new theories undermined classical Newtonian notions of strict causality and scientific determinism.

Einstein had fundamental qualms about abandoning strict causality and accepting that some things happened by chance. Repeatedly, he uttered his famous objection that he found the notion of God playing dice with the universe an incredible one. Quantum mechanics may be right, as far as it went, he admitted, but it did not seem complete. There must be, he thought, some deeper field theory that could be discovered, a unified theory that would tie together all of the forces of nature, such as electromagnetism and gravity, and in the process restore certainty and classical determinism to the laws of the universe.

Einstein had been among the first scientists to realize—and to worry—that the quantum nature of radiation might infect the cosmos with uncertainty, probability, and chance. He expressed it in a 1917 paper, "On the Quantum Theory of Radiation," that described how the emission of light quanta related to Niels Bohr's theory of how electrons in an atom change orbits in quantum leaps. Einstein described how this radiation could be stimulated artificially, which became the theoretical foundation for lasers. He also noted, less happily, that there was no way to determine which direction

ABOVE: *German physicist Max Planck. He is viewed as the father of quantum theory. Planck was also one of the few people who immediately understood the importance of Einstein's special theory of relativity.*

BELOW: *The Solvay Conference of 1911 met in Brussels to discuss "Radiation and the quanta" and the problems of having the two different approaches of classical physics and quantum theory. Einstein (seen here standing second from the right) was the youngest physicist to attend.*

an emitted photon might go or exactly when the emission would occur. While it was possible to calculate the probability that an atom would emit a photon at any specific moment, there could be no certainty about precisely when the photon would be emitted. Nor could its direction be predicted with absolute certainty. It did not matter how precise the measurements before the moment of emission, all that could be derived from them was a probability, just the sort of divine dice roll whose existence Einstein had so vigorously denied.

Whilst relativity had retained the idea of cause and effect, the quanta had about them a bizarre unpredictability which undermined this neat sense of causality. Einstein was forced to admit that there seemed a flaw in his theory. That it gave "chance" such a great role in such basic processes was hardly ideal. Einstein found the whole idea of simple chance—or "*Zufall*" as he expressed it, from the German—so distasteful that he put the word in quotation marks, as though seeking to gloss over its reality.

To some extent, his stubborn resistance to quantum mechanics, along with his futile search for a unified field theory, diminished the second half of his career. He seemed to sense this was happening. After finishing his work on general relativity, Einstein lamented to a friend that all really revolutionary thinking happened in youth, and, as one got older and more recognized in the field, stubbornness and the defense of one's own fixed notions displaced thinking about new ideas.

Some historians have said that science would not have suffered much if Einstein had retired after the eclipse observations and devoted himself to his hobby of sailing for the remaining 36 years of his life. There is some truth to this, but Einstein in fact played a useful role. Even though most of his attacks

Quantum Mechanics 97

on quantum mechanics did not prove to be warranted, he served to strengthen the theory by coming up with a few advances and also, less intentionally, by his ingenious but futile efforts to poke holes in it.

There is a broader question, however, that is raised by Einstein's resistance to quantum mechanics. Why was he so much more rebellious, and therefore more creative, before the age of 40 than he was after? Part of it is an occupational hazard of mathematicians and theoretical physicists, who historically have made their great breakthroughs before turning 40. The capacity to think naturally grows sluggish with advancing years and fame, though the aura of

NIELS BOHR
(1885–1962)

Born in Copenhagen, Bohr came from a family with an established scientific background, his father being a physiology professor at Copenhagen University. In 1910, he received his doctorate from Copenhagen for work on the behavior of electrons in metals. He then moved to Manchester, where in 1912 he worked in the laboratory of Ernest Rutherford, the discoverer of the atomic nucleus. Bohr published his atomic model in 1913, in which electrons were theorized as traveling in orbit around the atomic nucleus, and in 1916 was recognized by the Danish government with a professorship of physics at Copenhagen. His work on atomic structure was rewarded by the 1922 Nobel Prize for Physics. In 1941, he fled the Nazi occupation of Denmark, and after a stay in Sweden, moved to the United States, where he worked on the Manhattan Project to develop an American nuclear bomb.

LEFT: At the 1927 Solvay Conference Einstein's attack on quantum mechanics reached its peak. Each day Einstein would present Niels Bohr with thought experiments that he believed showed that quantum mechanics did not provide a complete description of reality. Bohr in return would work on disproving Einstein's thought experiments.

BELOW: While Einstein was busy professionally, he had found a new calm in his home life. This photograph, taken in March 1929, shows Einstein in his Berlin home with his wife Elsa (seated) and her daughter Margot.

scientific recognition still lends the illusion of respectability to the aging mind.

More specifically, there was a link between Einstein's creativity and his willingness to defy authority. He had no sentimental attachment to the old order, and thus was energized by upending it. His stubbornness and rebel instinct had worked to his advantage.

But by the time he reached 50, he had traded his youthful bohemian attitude for the comforts of a bourgeois home. Likewise, he had become wedded to his faith in the strict determinism of classical physics, and specifically to the idea that field theories could preserve that determinism. His stubbornness henceforth would work to his disadvantage. In one of his most revealing remarks about himself, Einstein later lamented how ironic it was that such a nonconformist had now himself become part of the establishment.

For the remainder of his life, Einstein remained opposed to the idea that the universe described by quantum mechanics was one in which uncertainty was the governing factor. He could not accept the notion that an electron could effectively "decide" in which direction and when to be emitted. If this were really so, Einstein remarked to physicist Max Born, then he had made a grave mistake in following a scientific career.

While visiting Berlin in 1920, Niels Bohr, who had become the standard-bearer of quantum mechanics, met Einstein for the first time. Bohr immediately launched into an exposition of the centrality of chance and probability in quantum mechanics. Einstein remained mistrustful, however, of completely giving up on continuity and causation. Bohr was not deterred and began to maintain that the abandonment of strict causality was so much supported by the evidence that there really was no other option.

The disagreement between the two men went to the fundamentals of how the cosmos was constructed. It concerned whether there was any objective reality, and whether, if there was, we would ever be able to observe it. It touched on whether the universe's course was predetermined or subject, in contrast, to total randomness and chance in its development.

Bohr would ever after regret his failure to convert Einstein to the quantum mechanics cause. Yet the discourse between them was civilized, and even at times humorous. When Einstein declared once more that God would not play dice with the universe, Bohr retorted that it would be as well if Einstein desisted from telling God what to do!

LEFT: *Berlin in the 1920s showing how it would have looked when Einstein lived there.*

ABOVE: *Einstein and Niels Bohr met at the home of Paul Ehrenfest in 1925 to discuss quantum mechanics. Photograph taken by Ehrenfest.*

WERNER HEISENBERG (1901–1976)

Born in Munich, Werner Heisenberg's father was a university professor of ancient languages, but Werner devoted himself to scientific studies, writing his doctoral thesis on turbulence. Heisenberg served as assistant to Niels Bohr in Copenhagen from 1926 to 1927, before becoming professor of theoretical physics at Leipzig in 1927. He published his theory of quantum mechanics in 1925, aged only 23. It was most memorably encapsulated in his "theory of uncertainty," which states in essence that we cannot ever know both the position and momentum of a subatomic particle with absolute certainty. His work was recognized with the Nobel Prize for Physics in 1932. He campaigned against Nazi mistreatment of Jewish and left-wing scientists and particularly against the leaders of the "German Physics" movement, but chose, unlike others in his position, to remain in Germany. In 1948, he became head of the Max Planck Institute and remained director there until 1970.

Einstein and Religion

When people heard Einstein invoke God, many assumed that it was purely a figure of speech, perhaps even an ironic one. He was, after all, not a religious person, despite the ethnic affinity he felt for the Jewish people. He never went to synagogue, and he clearly believed that nature's laws, rather than the hand of a wilful God, determined the daily workings of the cosmos.

Yet it was this beauty of the harmony of nature's laws, and a reverence for the mystery of creation, that made Einstein proclaim himself to be religious, in a sense, and to deny that he was an atheist.

At a dinner party in Berlin around the time Einstein turned 50, the conversation turned to astrology. Einstein dismissed it as superstitious nonsense. Another guest noted that religion was likewise just a superstition. But the hostess silenced the argument by noting that, despite his scientific views, Einstein himself had religious feelings.

"It isn't possible!" the skeptical guest said, turning to Einstein to ask if he was, in fact, religious. Einstein was calm and replied that behind all the physical laws with which the universe could be described, there was something else, elusive, indefinable, and subtle. Einstein's religion, as far as he had one, could be described as a reverence for this force.

For most of his life, Einstein had not said or written much about religion. He had briefly gone through an intense religious phase as a child, and then he had rebelled against organized religion around the time he developed an interest in science books. His enthusiasm for religion was apparently rekindled in the late 1920s, about the time of his 50th birthday, just when he began resisting the uncertainties inherent in quantum mechanics. As was evident on his trip to America, he had started to take a more profound interest in his Jewishness. He also, and to an extent independently, began to consider a belief in God, albeit a rather impersonal concept of God, with God as a force who had established the rules and order of the universe but who subsequently did not intervene in a direct fashion. This viewpoint is often known as "deism."

After an interview he gave to the journalist George Sylvester Viereck on religious matters, people began to bombard Einstein with further questions about his faith. So he composed a simple credo in the summer of 1930 that tried to give an elegant and simple answer. It was titled "What I Believe," and he both wrote it and made an audio recording of it. It concluded that a sense of mystery is the most profound of all emotions and lies at the basis of all real developments in the arts and science. Anyone who lacks this

RIGHT:
George Sylvester Viereck interviewed Albert Einstein in 1929 about his religion. Six years earlier, Viereck had interviewed Adolf Hitler.

VIERECK'S INTERVIEW WITH EINSTEIN

Einstein had just turned 50 when he gave one of his most revealing insights into his religious views. He was being interviewed by George Sylvester Viereck, a poet with a somewhat ingratiating nature and something of a talent as a propagandist. Einstein presumed that the German-born Viereck was Jewish, but in fact Viereck claimed a family connection with the German kaiser and would later become a supporter of the Nazis. He spent much of his life writing erotic poetry, interviewing men with far greater talent than his own, and penning complex justifications for his love of the German fatherland. Did Einstein, Viereck asked, consider himself to be a German or a Jew? Einstein responded that these were not mutually exclusive categories. Nationalism, he added, was a juvenile plague, the intellectual equivalent of measles. Quizzed as to whether Jews should assimilate, Einstein lamented that they had historically been too keen to sacrifice the differences that made them fundamentally Jewish. On his view of Christianity, Einstein explained that he had studied both Jewish and Christian scriptures and had a great deal of respect for Jesus Christ. The Gospels, he felt, were so imbued with the personality of Jesus that it was hard to doubt that such a person really existed. When the interview turned to whether Einstein believed in God, he denied being an atheist. The problem of God's existence, he clarified, was simply too great for human minds to contemplate. Just as a child who cannot read knows that a book contains great knowledge and that a library is organized according to some principle he cannot work out, so an intelligent person looking at the universe must understand there is some guiding force at work, but attempts to work out what this is or explain its characteristics will always be an oversimplification. On being asked if he believed in life after death, Einstein remarked, simply, that a single life span was quite enough for him.

GOLDSTEIN'S TELEGRAM

Around the time of Cardinal O'Connell's denunciation of Einstein's religious views, the Orthodox Jewish leader in New York, Rabbi Herbert Goldstein, sent Einstein a direct question by telegram: "Do you believe in God? Stop. Answer paid. 50 words." It took Einstein less than half that number of words to respond. He would later resort on many occasions to modified versions of this, his most famous answer to the question, in which he stated that he believed, like the Dutch philosopher Spinoza, in a God whose existence could be determined in the harmonious beauty of the natural laws he had established but who did not intervene in the day-to-day affairs of mankind. It was not an answer that pleased either side of the religious divide, but most people can probably understand, and even appreciate, what he was saying. The deist concept of a God who stands apart from the universe which he has created chimed with the personal beliefs of many of the founders of the United States, such as Thomas Jefferson and Benjamin Franklin, and it was reflected in the works of some of Einstein's favorite philosophers.

ABOVE: Rabbi Goldstein (center) in 1945 after attending an event at the White House. Goldstein believed Einstein's reply to his telegram was not only evidence that Einstein was not an atheist, but, if taken to its ultimate logical conclusion, Einstein's theory would lead to a scientific formula for monotheism.

sense, and who cannot feel transcendent awe at the wonder of the universe is, in effect, moribund. Einstein defined religion as the feeling that beyond even the marvelous experiences and emotions that touch us directly, there is a force that transcends it all. Only in his acknowledgment of this force could he be regarded as religious.

The statement became a sensation, and it was printed and distributed across the globe in a multiplicity of languages. Nevertheless, it did not settle the issue, which was hardly surprising. People kept posing questions to him about his religion in different ways. "Do scientists pray?" enquired a sixth-grader at a New York Sunday school. Einstein replied to the girl saying that the notion that there are natural laws which govern everything that happens is fundamental to scientific research. These laws should equally hold for the behavior of individuals. This is why scientists should be skeptical about the power of prayer: God would be hardly likely to break his own rules for an individual petitioner. But he followed that rejection of the concept of a personal God with an endorsement of the idea that there was, in fact, some force greater than us. He explained his ideas further to the girl, saying that all serious scientists eventually realize that there is a guiding hand in the laws of the universe, one whose power is so manifestly awesome as to leave us feeling humbled. It is this progression from science to a deeper understanding that imbues in scientists a religious sensibility quite different from the religious notions of everyday folk.

Throughout his life—and up until the current time—fragments of Einstein's various pronouncements were, and have been, embraced and denounced by both atheists and believers. Boston's Cardinal William Henry O'Connell was among those who denounced Einstein's denial of a personal God. "The outcome of this doubt and befogged speculation about time and space is a cloak beneath which hides the ghastly apparition of atheism."

Near the end of his life, Einstein wrote a letter, which was sold at auction with much publicity in 2008, debunking the idea of a personal God. He considered the word "God" as a mere summation of all that was weak about human nature and the Bible as just a compilation of childish myths, albeit ones whose longevity had imparted some worth to them. But throughout his life, Einstein consistently denied that he was a simple atheist. He was angered by those who said that God did not exist, but even more enraged by those who enlisted his words in support of their own views. In another letter, he noted that extreme atheists who have cast off organized religion are still in fact in thrall to their old beliefs. Their raucous rejection of hierarchy and dogma made them deaf to the real God, whose voice could be heard in the music of the sphere.

Most relevantly, Einstein did not feel that science and religion necessarily conflicted. In a talk at a conference on that topic at the Union Theological Seminary in New York, he argued that it was this idea of a personal God that was the main contemporary source of friction between science and religion.

His pithy conclusion became famous. He summarized his position on the arguments as that, devoid of religion, science could go nowhere, but bereft of science, religion was a blind guide stumbling in the darkness.

RIGHT: *Cardinal O'Connell dismissed Einstein's theories on religion for being "authentic atheism, even if camouflaged as cosmic pantheism."*

"The outcome of this doubt and befogged speculation about time and space is a cloak beneath which hides the ghastly apparition of atheism." —Cardinal William Henry O'Connell

Einstein and Religion

The Rise of Hitler

During the 1920s, Einstein's pacifism continued to grow deeper. When he attended a session of the League of Nations disarmament commission in 1928 that was looking into ways to curb the use of gas in warfare, Einstein did not hold back in expressing his contempt for such weak measures. He believed it was completely pointless to draw up rules and restrictions on how war should be conducted. War, he argued, was not a game, and so establishing rules for it was missing the point. The aim instead must be an end to war itself. He proposed setting up a body to encourage

BELOW: *It was rallies such as this one held by the Hitler Youth in the 1930s that encouraged Einstein to leave Germany in 1933.*

EINSTEIN CORRESPONDS WITH FREUD

In 1932, an antiwar group invited Einstein to exchange letters on the concept of pacifism with any prominent thinker he chose. Einstein's selection was Freud, the other great pacifist icon of the time, and the man who had an intellect to match his own. In his first letter, Einstein pushed the idea of world federalism that he had been thinking about for many years. He argued that in order to eliminate war completely, countries would have to cede some of their sovereign powers to a supranational organization. Specifically, this would have to be a body with far greater powers than the League of Nations. In a letter to Freud, Einstein wrote that a certain degree of surrender by countries of their individual sovereign rights would be the only way to safeguard the collective security of them all.

ABOVE: *Sigmund Freud, father of psychoanalysis. It seemed natural that the man with the greatest understanding of the universe should write to the man considered to have the greatest understanding of the human mind.*

young men to refuse service in the armed forces.

Pacifism was a growing movement at the time, a reaction to the meaningless horror of World War I. Einstein's compatriots in the cause included Upton Sinclair, Sigmund Freud, John Dewey, and H. G. Wells. "We believe that everybody who sincerely wants peace should demand the abolition of military training for youth," these luminaries declared in the Manifesto of the Joint Peace Council issued on October 12, 1930.

All Einstein's notions about world peace seemed to be rendered obsolete by Adolf Hitler's takeover in Germany in March 1933. Professional scientist that he was, however, Einstein was able to revise his theories when faced with new facts. Thus he revised his outlook toward pacifism.

When Hitler took power, Einstein was just completing a semester as a visiting scholar at the California Institute of Technology (Caltech) in Pasadena, and he would never go back to Germany again. As soon as his ship docked in Belgium, he disembarked and headed directly to the German consulate to turn in his passport and renounce his German citizenship.

He also renounced his pacifism. When a Dutch minister wrote to him asking for his support for peace, Einstein's reply was stark. In such times of crisis, radical pacifism was a dead end. He simply could not advise the refusal to serve in the military, a course which would render countries helpless in the face of a rapidly rearming Germany.

Einstein made his new views very public when he was called upon to help a powerful friend. He had made the acquaintance of King Albert I and Queen Elisabeth of

> **"We believe that everybody who sincerely wants peace should demand the abolition of military training for youth."** —Manifesto of the Joint Peace Council

ABOVE: King Albert I, his wife Elisabeth, and their son Leopold in 1905. Einstein's friendship with the Belgian queen was founded on a mutual love of music and the pair played together whenever possible.

Belgium, and he had occasionally joined the queen in playing the violin in an amateur string quartet. One day in 1933, when Einstein was living in temporary exile in Belgium, he got a cryptic message that he easily deciphered: "The husband of the second violinist would like to talk to you on an urgent matter." Einstein took himself off at once to the royal palace to see the king.

King Albert was worried that Einstein would accede to pressure being put on him by international pacifists to speak out on behalf of two conscientious objectors whose refusal to serve in the Belgian Army had led to their being jailed. Einstein agreed to the king's request that he support the Belgian government rather than the imprisoned pacifists. He made a public declaration that, in view of the present situation regarding Germany's build-up of its offensive capability, Belgium's armed forces should be seen principally as a means to deter aggression, not of promoting it.

He also wrote a public letter to the leader of the pacifist group in which he proclaimed that were he a young man, he would enlist in the army himself. If he were Belgian, Einstein wrote, he would consider it his duty to enlist in the defense of civilized values. *The New York Times* headlined the story: "Einstein Alters His Pacifist Views/Advises the Belgians to Arm Themselves Against the Threat of Germany."

Einstein had been offered a job at the new Institute for Advanced Study in Princeton, New Jersey and he now decided it was time to emigrate to America. He knew that his elder son, Hans Albert, would likely have the chance to follow him, as turned out to be the case. But his other son, Eduard, had succumbed to the mental demons that also plagued his mother Mileva and was now in an asylum in Zurich. Einstein knew that if he fled Europe, he might never see Eduard again.

Einstein's relationship with his former wife was now sufficiently cordial that she invited both him and Elsa to stay in her apartment building in Zurich. Pleasantly surprised, he took up her offer and stayed there in May 1933, on what would prove to be his last visit to Eduard.

EINSTEIN'S LAST VISIT TO EDUARD

Einstein's last visit to his younger son, Eduard, in May 1933 was a far more upsetting occasion than he had anticipated. Einstein had brought his violin with him, for on previous visits father and son had played a duet, finding expression in the music for the emotions which they could not find words to express to each other. A photograph of the visit survives, full of pathos. The pair are seated beside each other in what may be the asylum's visiting room. They seem ill at ease. Einstein has his violin and bow in his hands, and his glance is downcast, while Eduard seems more focused on a pile of papers than on his father. His now chubby face seems wracked with pain. Einstein could not be sure, but he probably suspected, that when he left for America later that year he would never return to Europe or see Eduard again.

RIGHT: *A rare picture of Einstein's two sons in July 1917 with Eduard on the left. Like his father, Eduard was a skilled musician, but unfortunately in 1930 he began to suffer from schizophrenia and had to be institutionalized for the first of many times in 1932. Following his diagnosis, his relationship with his father broke down. He died in 1948.*

To America

Albert Einstein arrived in the United States, aged 54, in October 1933, and he would live there for the remaining 22 years of his life. Indeed, from that point on, he would rarely spend a night away from Princeton, New Jersey, where he had been recruited to work at the Institute for Advanced Study by its director, Abraham Flexner.

Flexner showed concern for Einstein's privacy. He arranged for a tug to remove Einstein quietly away from the ocean liner that had brought him from Belgium as soon as it had cleared customs, "All Dr. Einstein wants is to be left in peace and quiet," he told to the reporters. Einstein also wanted an ice cream cone. So a few hours after he arrived in Princeton, he changed out of his formal clothes, and, puffing on his pipe, took a stroll down to a local ice cream parlor. A young divinity student was just being served, and Einstein gestured at the cone he had bought and then back at himself. As the star-struck waitress counted out his change, she declared: "This one goes in my memory book."

When shown his corner office at the Institute, Einstein was asked to indicate what he might need in the way of equipment. His requirements were not onerous, just a bit of office furniture and some stationery. What he wanted most of all was a large wastebasket, to which he could consign all his false starts and mistakes.

Once he and Elsa were settled into a new rental house, Einstein floated around Princeton with a distracted air and a wry sensibility. Thus arose the iconic image we have of Einstein in later life: the rather sweet and kindly old professor, roaming around distractedly, always ready to help out with a child's homework but utterly careless of his appearance, often even failing to wear socks. This he saw as some kind of rebellious act. He once confided to a neighbor that

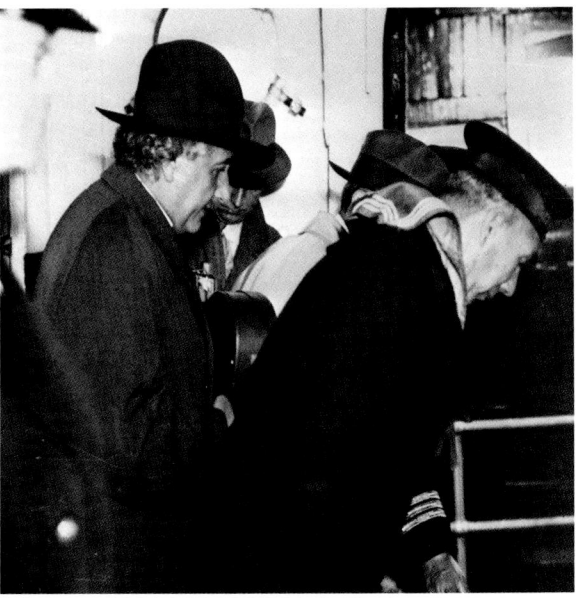

ABOVE: *Albert Einstein arriving in New York in 1933 after leaving Germany.*

> **"All Dr. Einstein wants is to be left in peace and quiet."**
> **—Abraham Flexner to reporters**

EINSTEIN AND MUSIC

Einstein soon became known around Princeton for his love of music and his genial demeanor. He and Elsa held a little welcome party that featured a musical recital. The talented Russian violinist Toscha Seidel took the lead, with Einstein accompanying him. Seidel gave Einstein some violin tips, and Einstein tried in turn to elucidate his relativity theory. He even drew some diagrams, which the violinist saved as a prized memento. On one occasion Einstein joined the virtuoso Fritz Kreisler's violin quartet. When they got out of sync at one point, Kriesler feigned exasperation. "What's the matter, professor, can't you count?" Another oft-told tale involved a Christian group that had gathered one evening to pray for persecuted Jews. They were delighted and surprised when Einstein asked if he could come. Producing his violin, he proceeded to play a solo, much in the manner of a prayer.

BELOW: *Einstein performing in a concert in aid of refugee children in January 1941. He was accompanied on the piano by concert pianist Gaby Casadesus.*

the best thing about reaching his relatively advanced age was that if someone told him to wear socks, he could just say no.

This impression of a rather rumpled genius was not far from the truth, but he enjoyed playing it to the hilt, since it was such a great role. He made it as famous as Charlie Chaplin's little tramp. Einstein's positive qualities were endearing; he was honest to a fault and tended to be almost naïve in his passionate commitment to humanity's welfare, and, occasionally, to that of individuals. His rather distracted gaze seemed constantly fixed on higher truths about the cosmos, a preoccupation which made his apparent detachment from more mundane concerns all the more comprehensible.

A woman he knew introduced him to cotton sweatshirts bought at an army surplus store, and he favored them over sweaters because of his mild allergy to wool. He had never been a stickler for haircuts and hairbrushes, and now his haphazard grooming became another quirk that he enjoyed. Likewise, all three of the women who had moved in with him—his wife Elsa, her daughter Margot, and his sister Maja—sported the same disheveled profusion of gray hair.

Elsa loved Princeton and did not want to leave. "The whole of Princeton is one great park with wonderful trees," she wrote to a friend. "We might almost believe that we are in Oxford." She did however suffer pangs of guilt that she and her husband lived so well, while others in Europe experienced terrible deprivation or persecution. "We are very happy here, perhaps too happy. Sometimes

ABOVE: *When he met Einstein in 1932 Flexner was developing the idea for a "haven" where academics would be able to work without academic pressures or the need to teach. He realized that Einstein would fit in perfectly at what was to become the Institute for Advanced Studies in Princeton.*

RIGHT: *Einstein's study at the Institute for Advanced Study in Princeton.*

ABOVE: *Einstein's home on Mercer Street, Princeton, New Jersey.*

BELOW: *Einstein working in the garden with his secretary Helen Dukas, who worked for him from 1928 until his death. After Elsa died, Dukas also assumed the role of housekeeper.*

one has a bad conscience." Their original plan had been to return to Europe for a semester, where Einstein would teach at Oxford or one of the other universities that had offered him a job. But by April 1934, they decided that they had no desire ever to see Europe again.

Einstein became a well-known fixture of Princeton life, often to be observed ambling along lost in thought, and years later almost anyone who lived there would tell tales of watching him. Because he was too distracted to be a good driver—"it's too complicated for him," Elsa often said—Einstein would make his way to the Institute each morning on foot. He would return for lunch at home each day, often joined on the walk by a small group of his professorial colleagues or by students. Einstein would seem in a dream, whilst the others tried, often in vain, to attract his attention with exaggerated gestures and loud voices. When they arrived at his house on Mercer Street, Einstein would often just stand there, still captivated by some thought or other, while the others, in despair, would drift off to their own homes. So distracted was he, that sometimes he would forget where he was going and start to shuffle back to the Institute. His secretary Helen Dukas, who was always on the lookout for him, would gently take his arm and lead him back inside, where she would make sure he was served macaroni for lunch. After eating and a quick nap, he would dictate replies to letters he had received, and then make his way to his study for a few hours' further consideration about the unified field theories which so preoccupied the latter part of his scientific life.

As he made his apparently aimless way around town, sometimes Einstein would become completely lost. One day, the secretary of the Institute answered the telephone. The caller said he needed to speak to one of the deans. When she replied that he was unavailable, the caller asked for Einstein's address. She replied that it was confidential, and she could not possibly give it out. In an embarrassed whisper, the caller then confessed that he was Einstein and that he had genuinely forgotten where his house was. He pleaded with the secretary not to let on about his little lapse of memory.

This idyllic life was jolted when Elsa became afflicted with heart and kidney problems. The doctors ordered her a strict regime of bed rest. On many evenings, Einstein would read to her, but generally he coped with the situation by immersing himself ever more deeply in his studies.

When she died in December 1936, Einstein cried, just as he had done on the death of his mother. His grief was greater than he had expected. Their relationship had not been a greatly romantic one. Except when he was courting her, Einstein's letters to Elsa have few sweet endearments. He was tetchy and trying towards her, and he showed little concern for her emotional needs. He had even been known to act flirtatiously toward other women.

However, there was a depth to their relationship. Even if it was not high romance, there were genuine and solid feelings of affection between them. The strength of their marriage depended on their understanding and responding to the other's needs and desires, providing a satisfaction which each of them valued. The depth of their insight, coupled with a genuine liking, was capped with a mutual appreciation of each other's humor (for Elsa, in her own way, was possessed of a genuine wit).

Einstein's way of avoiding the emotional pain of her death was, not surprisingly, to throw himself into his work. He was finding it hard to concentrate, he told his son Hans Albert, but work was the only way he could find to escape the all-too-personal feelings of grief. As he had done before at times of personal crisis, he buried his emotional pain in his dedication to his work, the thing which most gave meaning to his life.

"We are very happy here, perhaps too happy. Sometimes one has a bad conscience."
—Elsa Einstein on Princeton

LEFT: *From left: J. Robert Oppenheimer, Elsa Einstein, Einstein, Margarita Konenkov, and Elsa's daughter Margot in 1935, the year before Elsa died.*

RIGHT: *Einstein in his study at Mercer Street in June 1938.*

EINSTEIN'S HOUSE ON MERCER STREET

Soon after arriving in America, the Einsteins bought a white clapboard house very close to where they had been renting. It had a little front yard and porch facing on to a tree-lined street. This new home, 112 Mercer Street, seemed a perfect mirror of the famous man who now lived there. It was modest but charming, without airs and graces, and yet, situated as it was on a main road, it was both highly visible and somewhat secretive behind its hedge and veranda. The one renovation was building an office for Einstein on the second floor. Elsa oversaw the project. A new picture window looked out on the backyard garden. There were floor-to-ceiling bookcases and a table for Einstein that became permanently awash with his papers, pencils, and pipes. Einstein would sit on an easy chair in the room with a notepad on his lap, scribbling equations and gazing out of the window.

The Bomb

The rise of the Nazis caused Einstein to abandon his pacifism and take up a new cause when he settled in America: the effort to help Jewish refugees. Among those he assisted was Leó Szilárd, a physicist from Hungary who was an old acquaintance, with an air of eccentricity and charm almost to match Einstein's own. In the past Einstein and Szilárd had worked together on a new sort of refrigerator.

When he was forced to flee the Nazis, Szilárd had found himself in London, where, while waiting at a traffic light, he had conceived of the possibility of creating a nuclear chain reaction. By 1939, he was working on this topic at Columbia University, and when he heard about the discovery of fission using uranium, he realized that element might be used to produce this phenomenon.

Szilárd was deeply worried that the German government might try to purchase all the uranium from the Congo, then a Belgian colony. He mentioned his concerns to his friend and fellow-exile from Hungary, the physicist Eugene Wigner, and they decided that they had to find a way to warn

ABOVE: *Leó Szilárd hoped that the threat of the U.S. having a nuclear bomb would be used to force Germany and Japan to surrender.*

BELOW: *Charles Linbergh, left and Luftwaffe commander Hermann Göring (center) at a reception in Germany in July 1936.*

CHARLES LINDBERGH (1902–1974)

Charles Lindbergh achieved celebrity status after his solo flight across the Atlantic in 1927. But by 1939 Lindbergh had been spending a lot of time in Germany and had even been awarded that country's medal of honor by the Nazi Hermann Göring. He was tending towards isolationism, and he and Roosevelt did not see eye to eye politically. Soon after being contacted by Einstein, Lindbergh gave a nationwide radio address that was a clarion call for isolationism. "The destiny of this country does not call for our involvement in European wars," Lindbergh began. His speech was laced with lightly veiled pro-German comments and barbs about Jewish domination of the American media. Szilárd's next note to Einstein contained a wry understatement. "Lindbergh is not our man," he wrote.

the Belgians. So Szilárd decided to contact Einstein, who was, the Hungarians knew, on friendly terms with the Belgian queen mother.

On Sunday July 16, 1939, Wigner drove Szilárd out to the town of Peconic in northeastern Long Island, where Einstein was renting a cottage for the summer. As Einstein sat out with his guests on the porch around a plain wooden table, he listened intently to Szilárd's ideas about how graphite-coated uranium might produce a chain reaction. Einstein was astonished; it was a development he had never considered. After a few questions to get more clarification, he puzzled over the idea for a further quarter of an hour before grasping the awesome implications of Szilárd's insight. Einstein then put forward the idea, that, instead of writing to the queen mother, they might instead get in touch with a Belgian minister with whom he was acquainted.

Wigner was understandably cautious about a group of refugees contacting a foreign government without the knowledge of the State Department. So he suggested that the proper approach might be a letter from Einstein to the Belgian ambassador, with a cover letter to the State Department. Having agreed on an approach, Einstein then dictated a letter in German, which Wigner translated, before giving it to his secretary to be typed and then passed on to Szilárd.

Szilárd then mentioned the plan to Alexander Sachs, a Lehman Brothers economist, who happed to be a friend of President Roosevelt. Rather more worldly-wise than the trio of theoretical physicists, he offered to deliver the letter directly to the White House. Einstein liked the idea and invited Szilárd to come back out to Peconic so that they could revise the letter.

This time Szilárd enlisted as a driver Edward Teller, yet another Hungarian refugee who

ABOVE: *Einstein in 1939.*

LEFT: *Physicist Eugene Wigner has been labeled by some commentators as an intellectual equal to Einstein, albeit without the fame.*

was a theoretical physicist. Einstein was well aware that they were now making a far more serious approach than the initial plan of warning off Belgian ministers from supplying Congolese uranium to the Germans. Instead, he was planning to tell the United States president to consider the construction of an unimaginably destructive atomic weapon. So he dictated a draft of an entirely new letter. "Einstein dictated a letter in German," Szilárd recalled, "which Teller took down, and I used this German text as a guide in preparing two drafts of a letter to the president."

In his draft, Einstein explained that chain reactions were a theoretical possibility but with a potential practical application in building a new type of bomb. He urged the president to set up a working group of scientists to examine the possibility. On the basis of this draft, Szilárd composed a formal letter which Einstein signed.

Their odd initial decision to use the aviator Charles Lindbergh as a conduit to the president had to be revised once it was clear he was an isolationist with crypto-German sympathies. So they went back to the plan of depending on the financier Alexander Sachs. Even though the letter that they gave him was so clearly important, Sachs did not find an occasion to deliver it for nearly two months. By then, its urgency was even more obvious. At the end of August 1939, the Nazis and Soviets stunned the world by signing their war-alliance pact (also known as the Molotov–Ribbentrop Pact), and the next month proceeded to carve up Poland.

LEFT: *Alexander Sachs, the man who took Einstein's message to President Roosevelt.*

EINSTEIN'S FBI DOSSIER

By 1939, J. Edgar Hoover had been director of the FBI for 16 years, a post he would continue to hold for a further 32 years. He provided a dossier to the army that pointed to Einstein's pacifism and socialism as reasons to deny him a security clearance. The conclusion was stark. "In view of this radical background, this office would not recommend the employment of Dr. Einstein on matters of a secret nature … it seems unlikely that a man of his background could, in such a short time, become a loyal American citizen." So it was that the greatest scientific genius in the United States was denied any active knowledge of the country's greatest scientific project.

RIGHT: *J. Edgar Hoover sitting in his office at the FBI in 1939. Hoover began compiling a file on Einstein when he arrived in the U.S. in 1933.*

BELOW: *In September 1939 President Roosevelt gave a radio address stating that the U.S. would remain neutral in World War II. When he received Einstein's letter a month later, however, he felt compelled to act.*

Sachs finally got an appointment with President Roosevelt on the afternoon of Wednesday, October 11. He brought with him Einstein's letter and an 800-word summary of it he had made. He worried that if he left the letter with Roosevelt, the president might just look at it, set it aside and forget the matter. So he insisted on reading it aloud along with his memo. "Alex, what you are after is to see that the Nazis don't blow us up," the president said. "Precisely," Sachs replied. Roosevelt summoned his personal assistant. "This requires action," he declared.

A committee was immediately established with the task of coordinating the scientific knowledge about a bomb. When it met in Washington in October, Einstein was not there. Ironically, despite the fact that he had been so instrumental in alerting Roosevelt to the dangers of not pursuing a bomb, Einstein himself was perceived as a security risk, and not trustworthy enough to pursue the work himself.

ABOVE, RIGHT: *A group of Jewish refugees from Austria wave excitedly as they arrive in America during World War II. Einstein worked hard to help refugees such as them.*

EINSTEIN'S ASSISTANCE TO REFUGEES

Einstein pursued his attempts to help refugees both in public and in private. As well as frequent appearances at fundraising dinners and speeches, from time to time he gave violin recitals. In order to encourage donations, the organizers would ask guests to make out their checks to Einstein himself. He in turn would endorse them to Jewish or refugee relief charities, and when the banks returned the canceled checks to the original signatory, the donor would have as a souvenir the great man's autograph on their check. He also sponsored dozens of would-be migrants who needed a financial guarantee, particularly necessary, as countries such as the United States were increasingly reluctant to grant visas, while the British had cut down on permitted emigration to Palestine. In particular, he helped fellow Jewish scientists who were flocking to America as fast as they could be allowed in.

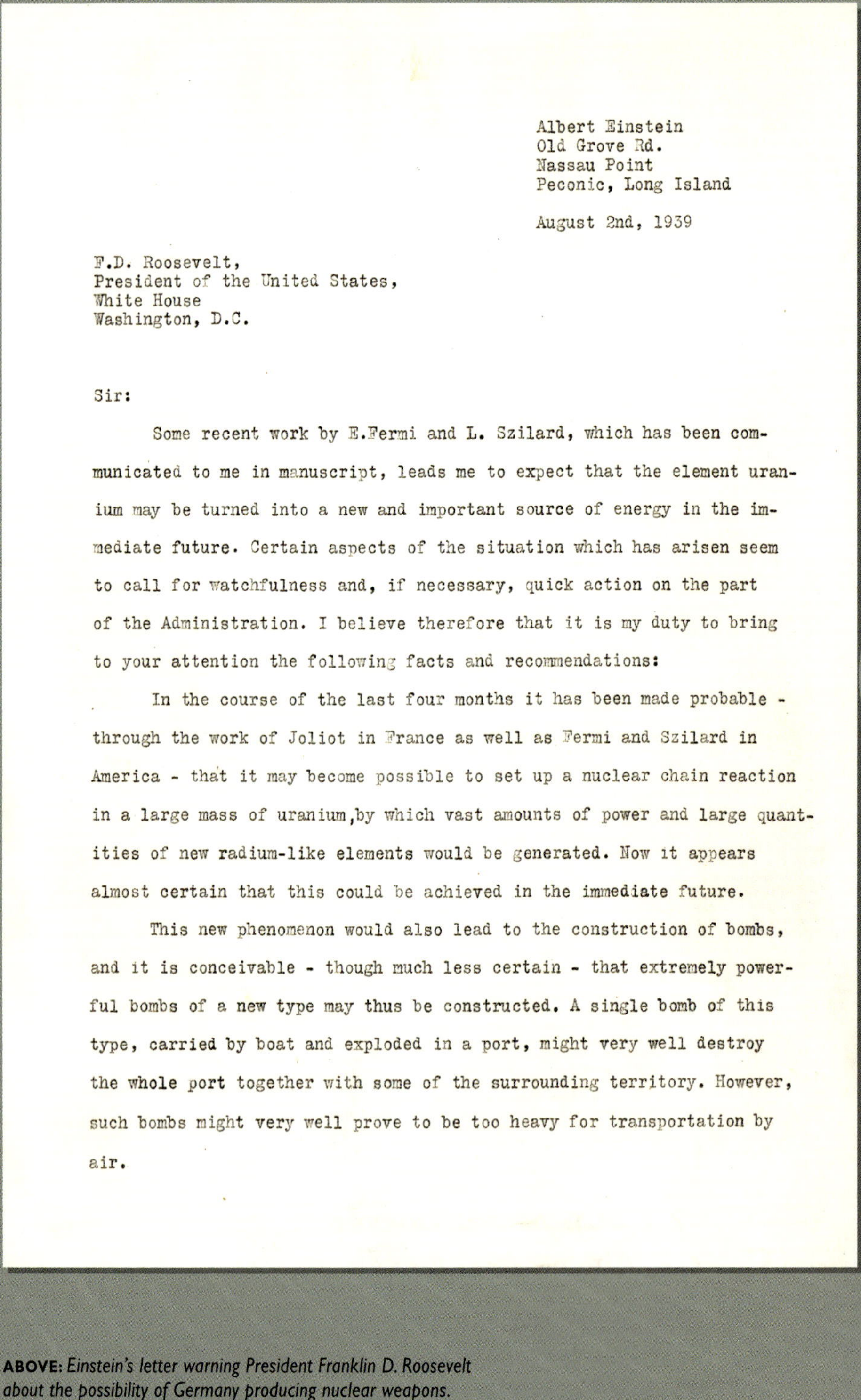

Albert Einstein
Old Grove Rd.
Nassau Point
Peconic, Long Island

August 2nd, 1939

F.D. Roosevelt,
President of the United States,
White House
Washington, D.C.

Sir:

Some recent work by E. Fermi and L. Szilard, which has been communicated to me in manuscript, leads me to expect that the element uranium may be turned into a new and important source of energy in the immediate future. Certain aspects of the situation which has arisen seem to call for watchfulness and, if necessary, quick action on the part of the Administration. I believe therefore that it is my duty to bring to your attention the following facts and recommendations:

In the course of the last four months it has been made probable - through the work of Joliot in France as well as Fermi and Szilard in America - that it may become possible to set up a nuclear chain reaction in a large mass of uranium, by which vast amounts of power and large quantities of new radium-like elements would be generated. Now it appears almost certain that this could be achieved in the immediate future.

This new phenomenon would also lead to the construction of bombs, and it is conceivable - though much less certain - that extremely powerful bombs of a new type may thus be constructed. A single bomb of this type, carried by boat and exploded in a port, might very well destroy the whole port together with some of the surrounding territory. However, such bombs might very well prove to be too heavy for transportation by air.

ABOVE: *Einstein's letter warning President Franklin D. Roosevelt about the possibility of Germany producing nuclear weapons.*

-2-

The United States has only very poor ores of uranium in moderate quantities. There is some good ore in Canada and the former Czechoslovakia, while the most important source of uranium is Belgian Congo.

In view of this situation you may think it desirable to have some permanent contact maintained between the Administration and the group of physicists working on chain reactions in America. One possible way of achieving this might be for you to entrust with this task a person who has your confidence and who could perhaps serve in an inofficial capacity. His task might comprise the following:

a) to approach Government Departments, keep them informed of the further development, and put forward recommendations for Government action, giving particular attention to the problem of securing a supply of uranium ore for the United States;

b) to speed up the experimental work, which is at present being carried on within the limits of the budgets of University laboratories, by providing funds, if such funds be required, through his contacts with private persons who are willing to make contributions for this cause, and perhaps also by obtaining the co-operation of industrial laboratories which have the necessary equipment.

I understand that Germany has actually stopped the sale of uranium from the Czechoslovakian mines which she has taken over. That she should have taken such early action might perhaps be understood on the ground that the son of the German Under-Secretary of State, von Weizsäcker, is attached to the Kaiser-Wilhelm-Institut in Berlin where some of the American work on uranium is now being repeated.

Yours very truly,

A. Einstein

(Albert Einstein)

Arms Control

BELOW: *On August 6, 1945, Hiroshima became the first Japanese city to be hit by an atomic bomb. The bomb killed 70,000 to 80,000 people, injured another 70,000, and destroyed everything within a one-mile (1.6 km) radius.*

Even though he did not work on the Manhattan Project, Einstein is intimately associated in the public imagination with the atom bomb. A few months after the weapon was used against Japan in 1945, *Time* put him on its cover with a mushroom cloud behind him that had $E=mc^2$ emblazoned on it. *Newsweek*, likewise, did a cover on him, with the headline "The Man Who Started It All." This was a perception fostered by the U.S. government. It released an official history of the atom bomb project that assigned great weight to the letter Einstein had written to President Franklin D. Roosevelt warning of the destructive potential of an atomic chain reaction.

All of this troubled Einstein. He told *Newsweek* that, with hindsight, and knowing that Germany's atomic program was in fact doomed to failure, he would not have acted to alert Roosevelt about the need for America to develop its own atom bomb. His discomfort over the creation of the bomb—and the indirect role he played—did not cause Einstein to become a pacifist again. Instead, he dedicated himself more fully to the need to create a system of world federalism and the benefits thereof. Einstein argued that some form of global government was the only hope for rescuing humanity. He believed that sovereign states, left to their own devices, would continue to stockpile armaments, and that, inevitably, friction between them would escalate into world wars.

> **WILLIAM GOLDEN (1909-2007)**
>
> William Golden, who worked at the Atomic Energy Commission and had been tasked with the preparation of a report on arms control for Secretary of State George Marshall, visited Einstein in Princeton. The physicist insisted that Washington was not trying hard enough to bring the Soviet Union on board with its current plans for arms control. In his report on the meeting to Marshall, Golden commented that Einstein spoke "with almost childlike hope for salvation and without appearing to have thought through the details of his solution. It was surprising, though perhaps it should not have been, that, out of his métier of mathematics, he seemed naïve in the field of international politics. The man who popularized the concept of a fourth dimension could think in only two of them in considerations of world government."

Einstein henceforth had two passions, both of which reflected his belief that there should be an order which transcended lesser laws and considerations: his quest for a unified field theory that would bind together the forces of nature and his advocacy of a unified world governing system that would prevent nationalistic atomic competitions. His motivation for the latter quest was a sense of guilt about his role in the inception of the atom bomb project. At a Nobel Prize committee dinner in Manhattan, Einstein commented that the prize had been established by Alfred Nobel as a form of atonement for his creation of the most deadly explosives hitherto known and that he now found himself similarly in need of atonement. He considered that the physicists who had participated in the atom bomb projects bore a share of responsibility for its consequences, and should feel a collective sense of guilt.

Einstein's theory of world federalism was one in which there was a world "government" or "authority" which had a monopoly on the exercising of military force. He termed this a "supranational" rather than an "international" body, as it would act on a higher level to its constituent member nations rather than serve as a means of mediation between them. He explained his views in an exchange of letters he had with a news commentator named Raymond Gram Swing. The journalist used this to produce an article in the *Atlantic* magazine of November 1945 entitled "Atomic War or Peace."

The new world government, Einstein said in the article, should be created by the United States, Britain, and the USSR, who should in turn invite other countries to join it. The "secret of the bomb" (as it was then known in popular parlance) should then be entrusted by the United States government to the new world organization.

The onset of the cold war, however, made this difficult. By late 1945, the United States was confronting the Soviet Union over its imposition of pro-Soviet communist regimes in Poland and the other areas of eastern Europe which the Red Army had occupied at the end of World War II. The Soviet Union in turn saw itself as surrounded by hostile powers and sought a security buffer zone. Its leaders were paranoid about any external attempt to dabble in its domestic affairs, and this made them extremely suspicious about any suggestions about a world government which would dilute their authority at home.

Einstein tried to accommodate their objections. He explained that his world

LEFT: *Philippe Halsman's legendary photograph of Einstein.*

PHILIPPE HALSMAN'S PHOTOGRAPH OF EINSTEIN

Philippe Halsman was one of the most famous portrait photographers of the age and another Jewish refugee from Nazism whom Einstein intervened to help. In this case, it was because he was imprisoned in Austria, accused of having murdered his father, his sentence tinged with anti-Semitism. In 1947 he was carrying out a photo shoot of Einstein when he asked the scientist whether he thought there would ever be lasting peace in the world. Einstein responded that, wherever there were people, there would be war. At just that moment, Halsman took the photograph that has become one of the most enduring images of Einstein, his eyes imbued with a sad, knowing quality. It was this picture which would be used when Einstein's portrait was placed on a United States postage stamp in 1966.

RIGHT: *Einstein at a luncheon given by the Emergency Committee of Atomic Scientists. Front, from left: Harold C. Urey, Einstein, and Selig Hecht. Back, from left: Victor F. Weissko, Leó Szilárd, Hans A. Bethe, Thorfin R. Hogness, and Philip M. Morse.*

government was not intended to export Western style liberal democracy into the Soviet bloc. He did not consider that the three great powers needed to alter their constitutional structures, and that adherence to western notions of democracy was not a prerequisite for becoming a member of his hoped-for world security organization.

In order to further this idea, Einstein became the chairman of a new group called the Emergency Committee of Atomic Scientists, which dedicated itself to promoting both nuclear arms control and the ideals of a world government. Einstein explained that the power of the atom bomb had overturned all traditional military, political, and strategic thinking and that the possibility of a global catastrophe was now all too real.

Far from being naïve, as some American government officials thought, Einstein had a cold and hardened view of human nature, which came from having lived in Germany in the first half of the 20th century. That made him the ultimate realist. It was this assessment of human nature—rather than some woolly-headed naïveté—that led Einstein to advocate a world military authority. The only realistic alternative to a world government, he announced in 1948, was the total destruction of humanity. When he was asked what he thought the next world war would look like, he famously responded that he could not tell how World War III might be waged, but that the fourth would assuredly see a return to the weapons of Stone Age man.

If Einstein appeared to be naïve, it was only because he had no time for the compromises and half-measures associated with politics. Physicists do not modify their equations in order to get them accepted; they are either correct, or they are not. It is this black-and-white view of reality that makes scientists bad politicians. Einstein's advocacy of something totally new—a world government that had a monopoly on nuclear weapons—had similarities to his scientific breakthroughs. It involved being imaginative enough to abandon entrenched assumptions that others considered verities, but it did not take sufficiently into account the all too predictable resistance to his ideas by entrenched interests. Just as the absolute nature of time and space had been the bedrock of the scientific view of the cosmos for four centuries, so the sovereignty and military autonomy of the nation state had formed the foundation of the political order over that period. To propose sweeping all this aside seemed at first sight to be the product of a nonconformist genius, but, as with many of Einstein's notions, it would, had it been taken up, probably have come to seem less revolutionary once implemented.

Civil Rights

At the same time that the FBI was denying him security clearance, Einstein was acting in a way he had not done for four decades. Of his own volition, and with something of a sense of pride, he was applying to become a citizen of the United States. Although he no longer had a German passport, he had a Swiss one and Swiss citizenship, so he did not need to do this. But he had a strong wish to do so.

On June 22, 1940, Einstein appeared before a federal judge in Trenton for his citizenship test. He passed and on October 1 took the oath of American citizenship, together with 88 other new Americans, including his stepdaughter Margot and his assistant Helen Dukas. Speaking to reporters present at the ceremony, Einstein was effusive in his praise for his newly adopted country. The United States, he affirmed, would show that democracy was not merely a type of government, but that it was inextricably intertwined with the American tradition of moral strength.

What Einstein liked most about America—especially compared to Europe—was that it was, for the most part, free of rigid class hierarchies and distinctions. In his letters from Princeton when he first arrived, he marveled at this trait to friends back in England. As he grew to understand the United States more, Einstein came to value the toleration of free thought and free speech, and its acceptance of nonconformism, all characteristics which he found appealing. These traits had shaped his scientific thinking, and now they formed the foundation of his views on citizenship. He believed that it was the freedom of Americans to express themselves exactly as they chose that made life in the country so precious to them.

Ever since his student days in Switzerland, Einstein held firm to some basic premises in politics. He espoused a socialist economic policy, but his views were tempered by a passion for individual freedoms, strong democratic institutions, and the preservation of individual liberties. He counted as his friends many of the leading lights in democratic socialism in Britain and the United States, such as Bertrand Russell and Norman Thomas. In 1949, he set down his political views in an article for the first issue of the *Monthly Review*, which was entitled "Why Socialism?"

Capitalism, according to Einstein, caused, by its very nature, cycles of boom and bust and accentuated the disparity between rich and poor. Instead of promoting cooperation, it led to selfish behavior as a means of self-preservation. Its celebration of the acquisition of wealth as an end in itself and its discouragement of philanthropic behavior, he found uncongenial. Education in a capitalist system, he noted, tended to concentrate on careers rather than creative thought.

Einstein argued that all this could be avoided by replacing the current system with a socialist economy, as long as safeguards against over-centralization and the reduction of personal liberties were set in place. Einstein

EINSTEIN'S CITIZENSHIP

As part of the celebrations on his acquisition of American citizenship, Einstein consented to a slot on the "I Am an American" show, which was run by the immigration service. He was served lunch by the judge, as the radio production team made his chambers ready, everything designed to smooth the process for the scientific celebrity guest. Einstein gave an inspirational performance, a harbinger of the kind of forthright, free-speaking citizen he intended to be. Once more he argued, and this time to a much broader audience, that nations would have to cede a portion of their sovereignty to an international body with military powers. It would be pointless, he declared, to have a world military body which did not completely control the security forces of its members. Such an organization could never guarantee global peace.

BELOW: *Einstein taking the oath of allegiance to the United States in order to become a U.S. citizen. On his left is Helen Dukas and Margot is on his right.*

EINSTEIN'S HONORARY DEGREE

Einstein was offered a plethora of honorary degrees but rarely accepted them in person. He broke this rule, however, when he was offered one by Lincoln University, a black educational institution in Pennsylvania. He did not dress up for the occasion, characteristically sporting a tattered grey herringbone jacket. He gave a lecture for the students, going over his relativity equations on a blackboard. Afterward, in his graduation address, he fulminated against segregation as an unthinking acceptance of the prejudices of previous generations. To go against such behavior, he wanted to meet the university president Horace Bond's six-year-old son. The young boy, named Julian, would later go on to great things, becoming a state senator for Georgia, a leader of the civil rights movement, and the chairman of the NAACP (National Association for the Advancement of Colored People).

ABOVE: *Horace Mann Bond awarding Einstein his honorary degree from Lincoln University in May 1946.*

considered that the best way of assuring a decent living to all citizens was a planned economy in which work would be allocated to those best able and apt for it and in which production could be determined by what the community actually needed. Education in such a culture would, apart from imparting basic skills, focus on instilling a sense of responsibility toward fellow citizens instead of elevating the cult of money and power which, in Einstein's view, so afflicted contemporary society.

Einstein, however, added a rider to the effect that such planned economies had an inherent tendency to become bureaucratic and to squeeze out personal freedoms, precisely what had happened in the Soviet Union and other communist countries. He warned that planned economies could breed servility in the individual. It was vital, Einstein believed, for enlightened social democrats to address two key questions: how to stop bureaucracy from smothering individual initiative and how to ensure that personal rights were safeguarded.

Protecting the rights of the individual lay at the heart of Einstein's political philosophy. Without individual freedom, creativity could not flourish, and imaginative work in the arts and sciences would wither away. The idea of the state imposing restraints on the individual's freedom of action, whether personal, political, or professional, was repugnant to him.

This was manifest in his views on civil rights. He quickly became, after arriving in Princeton, an outspoken opponent of racial discrimination in America. It was a time when cinemas were still segregated and black people were excluded even from trying on shoes or clothes in department stores. The Princeton student newspaper still felt able to declare that open access for blacks to the university was "a noble sentiment but the time had not yet come."

As a Jew who had been raised in Germany, Einstein empathized deeply with those who were discriminated against in this manner. In an essay called "The Negro Question" for *Pageant* magazine, Einstein commented that the more he came to feel like an American, the more this belittling of fellow Americans distressed him. The only way he could escape a feeling of colluding in the situation was to speak out against it.

Einstein made many gestures that showed his distaste for segregation, including his decision to invite the black contralto Marian Anderson to stay with him, when, on a visit to Princeton for a concert, the local Nassau Inn refused her a room. Thereafter, Anderson would always stay with Einstein on her visits to Princeton, the last such occasion being just two months before his death.

BELOW: *Marian Anderson performing in 1947 in Carnegie Hall. Einstein and Anderson remained friends from 1937 until his death.*

RIGHT: Einstein's U.S. Certificate of Naturalization.

UNITED STATES OF AMERICA

No. 5013865

CERTIFICATE OF NATURALIZATION

...uralization: Age 61 years; sex Male ; color White ;
...Brown , color of hair Grey ; height 5 feet 7 inches;
...rks None
................................; former nationality German
...ven is true, and that the photograph affixed hereto is a likeness of me.

Albert Einstein
(Complete and true signature of holder)

...ted States of America } ss:
...trict of New Jersey }

Be it known, that ALBERT EINSTEIN
residing at 112 Mercer St., Princeton, New Jersey
...g petitioned to be admitted a citizen of the United States of America, and at
...n of the DISTRICT Court of THE UNITED STATES held pursuant to law at Trenton, New Jersey on October 1st, 1940
...urt having found that the petitioner intends to reside permanently in the
...d States, had in all respects complied with the Naturalization Laws of the United
...in such case applicable, and was entitled to be so admitted, the court thereupon
...ed that the petitioner be admitted as a citizen of the United States of America.
...n testimony whereof, the seal of the court is hereunto affixed this 1st
...f October in the year of our Lord nineteen hundred and
forty and of our Independence the one hundred
sixty-fifth.

Benjamin F. Havens
Clerk of the U. S. District Court.

By *Hazel K. Fris* Deputy Clerk.

The Endless Quest

Ever since the mid-1920s, Einstein had been lobbing thought experiments into the citadel of quantum mechanics. The notion that there was an inherent uncertainty in nature felt wrong to him. He assumed that there was an underlying reality, and he kept coming up with hypothetical experiments that could theoretically determine for certain how particles behaved. But Bohr and Heisenberg protected the edifice of quantum mechanics by finding flaws in Einstein's thought experiments.

One particular target Einstein kept aiming at was Heisenberg's uncertainty principle, which says that we cannot know for sure the

BELOW: Léon Rosenfeld (right) and German physicist Walter Heitler.

ABOVE: *From left: Vladimir Rojansky, Léon Brillouin, and Einstein's colleague Boris Podolsky.*

exact position and momentum of a particle at the same instant—and that the very act of observing the particle affected the certainty of its position. In 1933, just before he came to America, Einstein attended a lecture by a Belgian physicist named Léon Rosenfeld. Einstein asked a question, positing a situation in which two particles possessing large momentum move toward each other and interact at an identifiable position, but over a very short time span. After the particles have moved far away from each other, a measurement is taken of the momentum of one of them by an observer. Given the terms of reference of the experiment, an observer could work out the momentum of one particle from that of the other and, equally, by knowing where one particle was located, could calculate the position of the other. This showed, Einstein felt, that we could know the position and momentum of a particle without directly observing it. With the help of two associates at the Institute, Boris Podolsky and Nathan Rosen, Einstein did further work on this thought experiment. Together they wrote a paper, published in May 1935, which was known as the "EPR" paper. In their title they posed the question "Can the Quantum-Mechanical Description of Physical Reality be Regarded as Complete?"

Einstein and his coauthors expanded the previous thought experiment, in which the particles which collided are shown to have correlated properties. They claimed that the position of the second particle could be determined precisely by measuring that of the first, a principle which held similarly for measurements of the respective particles' momentum. From this they concluded that, because at any given moment we might measure the properties of the first particle and thus determine those of the second, then the position and momentum of the second particle, even though we have not observed it, must both be real. As quantum mechanics cannot account for the reality of these properties of particles, then the answer to the question they had set themselves in their title must be "No"— quantum mechanics cannot provide a complete description of reality.

The only way to refute this conclusion, the authors of the paper went on, would be to maintain that the very act of measuring the first particle would cause the position and momentum of the second particle to be affected. To maintain this, they concluded, was simply unreasonable.

Bohr, the leader of the quantum mechanics movement, had over the years rebutted Einstein's assaults, and his response to the latest attack was firm. Einstein's dismissal of Bohr's notion that the particles interacted, or were "entangled" and his belief that an observation of one particle cannot instantly affect the reality of a distant particle seems correct. Alas, physics is sometimes weirder than our intuitions. Over the years, scientists have found ways to detect the phenomenon,

The Endless Quest

which Einstein dismissively labeled as tantamount to paranormal activity.

Even though the physicist Erwin Schrödinger was a pioneer in developing quantum mechanics, he shared some of Einstein's qualms and was amongst those who hoped that Einstein would get the better of Bohr and his Copenhagen colleagues. When he read the EPR paper, he at once sent Einstein a congratulatory note. "You have publicly caught dogmatic quantum mechanics by its throat," he wrote. Both former rebels conceded that as they got older they became more conservative about clinging to the classical foundations of physics, such as strict causality and determinism. The easy morals of youth often turn to ultraconservatism in old age, Einstein wrote, and radical firebrands metamorphose into pillars of establishment.

The two scientific titans now began to work together on conceiving another thought experiment which might undermine the foundations of quantum mechanics. They focused on what would happen when the quantum world, which included subatomic particles, interacted with the macro-world of our daily life.

In the quantum sphere, a particle cannot be definitely located at any given time. All that can be ascertained is a wave function, a mathematical description of the probability of finding the particle in any given place. These wave functions are also necessary to describe "quantum states" such as the probability of whether an atom will have decayed or not when it is observed. According to the orthodoxy espoused by the Copenhagen quantum mechanics school, the only reality—until it is observed—that a particle possesses, as far as its position or state goes, is precisely these probabilities.

Einstein suggested a clever thought experiment to Schrödinger. A pile of gunpowder might combust at any time because a particle within it is unstable. Quantum mechanics would describe this as a mixture of a system that had already blown up and one on the point of combustion. But this, in reality, is clearly impossible, for a thing has either exploded or it has not.

Schrödinger came up with a more vivid version of the thought experiment that was intended to show the bizarreness of the indeterminacy which is caused when the quantum realm interacted with the world of our senses. He imagined what would happen to a hypothetical cat. It became known as Schrödinger's cat:

> "A cat is penned up in a steel chamber, along with the following device: in a Geiger counter there is a tiny bit of radioactive substance, so small, that perhaps in the course of the hour one of the atoms decays, but also, with equal probability, perhaps none; if it happens, the counter tube discharges and through a relay releases a hammer which shatters a small flask of hydrocyanic acid. If one has left this entire system to itself for an hour, one would say that the cat still lives if meanwhile no atom has decayed. The psi-function of the entire system would express this by having in it the living and dead cat (pardon the expression) mixed or smeared out."

Einstein thought Schrödinger had scored a direct hit at the core of quantum mechanics. A wave function that incorporated both a live and a dead cat was obviously not a description of any possible reality. But adherents have made an industry of saving this poor cat from imaginary suspended animation. Even the cleverness of Einstein and Schrödinger combined did not prove quantum mechanics wrong.

Nor did Einstein make much progress on his goal of finding a unified theory that would tie together the forces of nature, reconcile quantum theory with relativity theory, and dispel all the uncertainty that seemed inherent in quantum mechanics. Yet even as he went into semiretirement in Princeton, he ambled to the Institute each day to continue this quest.

Einstein often sent drafts of each stab at a unified theory to Schrödinger, whom he believed to be the only one of his scientific acquaintances who was not blindly wedded to indeterminacy. But each new approach ultimately led to dead ends. He complained to Schrödinger that he had spent an inordinate amount of time on one of them, only to produce a diabolical mess.

Yet Einstein pressed on, producing a mass of papers on the subject, and occasionally coming to the attention of headline writers. When he published a version of one of his papers, the *New York Times* dedicated its front-page to his dense equations, together with the headline "New Einstein Theory Gives a Master Key to Universe; Scientist, After 30 Years' Work, Evolves Concept That Promises to Bridge Gap Between the Star and the Atom." But even Einstein realized that the promise remained an unfulfilled one.

Despite 30 years of searching, Einstein never came up with anything tangible that furthered the state of physics. He never hit on an insight that would enable him to crack the code, nor conjured up a thought experiment that would go to the heart of the problem. It may have been a fool's errand, but, should a unified field theory ever be discovered, then Einstein's persistence will come to seem a great deal less quixotic. Einstein would never regret the huge effort he had put into his quest. When one of his colleagues chided him for the amount of time he was putting into this clearly futile endeavor, he simply replied that he believed his effort was worthwhile. His reputation as a scientist was secured, and so he had no need to achieve instant results and could afford to dedicate himself to a cause which might find no resolution. A younger man might not be able to do so, since such doggedness might ruin a less-established career. Einstein, therefore, whatever the apparent purposelessness of his work, felt it his duty to press on.

ERWIN SCHRÖDINGER (1887–1961)

Vienna-born, Erwin Schrödinger studied theoretical physics at his home city's university, although his academic career was interrupted by a stint serving as an Austrian artillery office in World War I. After short postings at various German universities, he became professor of physics at Zurich in 1921, a position he held until his move to Berlin in 1927. In Switzerland, Schrödinger developed the key equation—which would be named after him—which describes how the quantum state of a physical system changes over time, work for which he would be awarded the Nobel Prize in 1933. He became an outspoken critic of Nazi policies and was forced to return to Austria in 1933, but the German occupation of his country in 1936 led to his flight to Ireland, where he remained for 17 years. In his later career, Schrödinger shied away from some of the developments to which his own work had opened the way, and he was an ardent supporter of Einstein in seeking to undermine Niels Bohr and the hardline supporters of quantum mechanics.

Israel

Einstein was not originally in favor of the creation of the State of Israel. Although Einstein did support Jewish migration to Palestine, he did not believe in a Jewish nation-state, because the whole idea of nationalism went against his world federalist sentiments.

Einstein was asked to testify to an international committee that met in Washington after the war to look into the situation of Jews in Palestine. Zionists in favor of a Jewish nation-state hoped that by then the horrors of the Holocaust would have caused him to voice support for their cause. He did not do so. Instead, he called for more Jewish immigration and blamed the British for stoking up animosity between Jews and Arabs. But in a quiet voice that dismayed the ardent Zionists in the room, he declared that

ABOVE: *Einstein speaking in 1946 before the Anglo-American Committee investigating the situation in Palestine.*

BELOW: *The front page of the* New York Times *on May 15, 1948 announcing the creation of the State of Israel.*

EINSTEIN'S SPEECH AT THE MANHATTAN SEDER

One of Einstein's best expressions of his opposition to a Jewish state came before the outbreak of World War II, as he was speaking to 3,000 celebrants at a Manhattan hotel Seder (the Passover dinner.) He explained that his idea of a Jewish State did not involve armies, impermeable frontiers, and the enforcement of laws by a powerful government. He felt this would actually damage what was precious about Judaism. Whilst Jews during the Maccabean period of the first and second centuries BC might have needed a narrow sense of nationalism to survive, these were sentiments no longer appropriate to modern times.

DAVID BEN-GURION
(1886–1973)

Born in Tsarist Poland, David Ben-Gurion (or Grün as he was originally named) settled in Ottoman Turkish-held Palestine before his expulsion to the United States in 1915. By now an ardent Zionist, he returned to Palestine after World War I and worked fervently during the period of the British mandate for the establishment of a Jewish state. It was Ben-Gurion who declared the State of Israel, which was established on May 14, 1948, and he became its first prime minister. When Weizmann died and the name of Einstein came up as a potential president, Ben-Gurion seemed to realize that the scientist was not cut for the cloth of the presidency. He was secretly relieved when Einstein declined. Half-jokingly, he confided to his assistant that he had been rather afraid the scientist might say yes. Offering the presidency to Einstein had become an inevitable gesture, but his acceptance of it would have caused an equally impossible situation. Ben-Gurion himself remained prime minister until 1963 and led his country through the turmoil of the 1956 Suez Canal Crisis.

he did not feel an emotional attachment to the proposal for a Jewish state, the need for which he doubted.

Einstein finally changed his position when the State of Israel was declared in 1948. Once again, as a scientist he was willing to revise his theories based on new facts. He wrote a friend explaining that although he saw no compelling economic, political, or military arguments in favor of a Jewish state, he felt the momentum for its establishment was now unstoppable.

His friend Chaim Weizmann, who had brought him to America for the triumphal tour of 1921, became Israel's first president. Because most of the power in the parliamentary system was vested in the prime minister, the post was mainly ceremonial. When Weizmann died in November 1952, press and public sentiment began to build for the idea of recruiting Einstein to succeed him. Prime Minister David Ben-Gurion, with some trepidation, decided he had little choice but to make such an offer.

When he read in an article in the *New York Times* about the possibility that he would be offered the presidency, Einstein initially thought it was a joke that would go away. He and his secretary Helen Dukas, along with his sister and stepdaughter, played a game of coming up with silly ideas about what he could do as president and the people he could appoint. But he began to take things more seriously when the news reports persisted and a telegram arrived from Israel's ambassador in Washington, Abba Eban, asking for a meeting. Einstein lamented to a visitor that this put him in an awkward position. To Dukas he complained that it was pointless for Eban to travel so far when the response would inevitably be negative.

It was the days before dialing long distance was common, especially among older people. So it was a small brainstorm when Dukas came up with the idea of simply picking up the phone and calling Ambassador Eban. She was astonished to be able to track down Eban and have him speak directly to Einstein, who then explained to the ambassador that he was simply the wrong person for the job and so would have to decline the post.

"I cannot tell my government that you phoned me and said no," Eban replied. "I have to go through the motions and present the offer officially."

So Eban sent a deputy to Princeton with an official letter offering the presidency. Eban's letter promised Einstein that there would not be much work and that the "freedom to pursue your great scientific work would be afforded by a government and people who are fully conscious of the supreme significance of your labors." Eban did feel compelled to note, presumably in case Einstein thought he could do the job from a distance, "acceptance would entail moving to Israel and taking its citizenship."

The offer really was a wonderful acknowledgment of Einstein's standing as a hero—perhaps the most famous and beloved of all—among Jews worldwide. It "embodies the deepest respect which the Jewish people can repose in any of its sons," Eban said.

Einstein immediately handed his letter of rejection to Eban's envoy. In it he declared although he found the honor deeply moving, he could not accept it. Einstein said he felt ashamed not to be able to accept as his relationship with his fellow Jews was strong, but his life had taught him to be objective, leaving him to feel he lacked the ability and experience necessary to deal with people and to assume an official role.

Einstein had lived long enough to know that not every clever idea was a good one, and the idea of having him as Israel's president certainly fell into the category of clever but misguided. He was not a natural at playing ceremonial roles, dealing with pomp, or soothing the egos of contending people. Though friendly and often genial, he was never particularly collegial. He was impatient with the compromises that marked out a good manager, or even a figurehead leader, of a complex organization. He liked to say precisely what he believed, rather than cloak his thoughts in diplomatic language, and he was not cut out to be either a statesman or a figurehead.

What made Einstein a brilliant scientist was what also made him an unlikely politician: he relished being a rebel and nonconformist who recoiled at any attempt to restrain his free expression. Although he admitted to a friend that many rebels end up by being respected and responsible, this was not a road down which he himself wished to travel. As he explained in a letter to a Jerusalem newspaper, he would not put himself in the position of having to stay silent about a government decision that might embark on policies opposed to his own moral values.

BELOW: *Israel's ambassador in Washington, Abba Eban.*

EMBASSY OF ISRAEL
WASHINGTON, D.C.

שגרירות ישראל
ושינגטון

November 17, 1952

Dear Professor Einstein:

        The bearer of this letter is Mr. David Goitein of Jerusalem who is now serving as Minister at our Embassy in Washington. He is bringing you the question which Prime Minister Ben Gurion asked me to convey to you, namely, whether you would accept the Presidency of Israel if it were offered you by a vote of the Knesset. Acceptance would entail moving to Israel and taking its citizenship. The Prime Minister assures me that in such circumstances complete facility and freedom to pursue your great scientific work would be afforded by a government and people who are fully conscious of the supreme significance of your labors.

        Mr. Goitein will be able to give you any information that you may desire on the implications of the Prime Minister's question.

        Whatever your inclination or decision may be, I should be deeply grateful for an opportunity to speak with you again within the next day or two at any place convenient for you. I understand the anxieties and doubts which you expressed to me this evening. On the other hand, whatever your answer, I am anxious for you to feel that the Prime Minister's question embodies the deepest respect which the Jewish people can repose in any of its sons. To this element of personal regard, we add the sentiment that Israel is a small State in its physical dimensions, but can rise to the level of greatness in the measure that it exemplifies the most elevated spiritual and intellectual traditions which the Jewish people has established through its best minds and hearts both in antiquity and in modern times. Our first President, as you know, taught us to see our destiny in these great perspectives, as you yourself have often exhorted us to do.

        Therefore, whatever your response to this question, I hope that you will think generously of those who have asked it, and will commend the high purposes and motives which prompted them to think of you at this solemn hour in our people's history.

        With cordial personal wishes,

Yours respectfully,

Abba Eban

Professor Albert Einstein
Princeton, N.J.

ABOVE: *Abba Eban's letter asking Einstein to assume the presidency of Israel.*

Red Scare

Despite the fact that J. Edgar Hoover's FBI had balked at giving him a security clearance, Einstein was a good and proud American. Admittedly, he was a somewhat contrarian citizen. But in that regard he was following the tradition of some venerable strands in the fabric of the American character: fiercely protective of individual liberties, often cranky about government interference, distrustful of great concentrations of wealth, and a believer in the idealistic internationalism that gained favor among American intellectuals after both of the great wars of the last century.

He had forsaken Nazi Germany with his public pronouncement that he would not live in a country where people were denied the freedom to hold and express their own thoughts. He later wrote that he failed to appreciate at the time how right he had been to come to live in the United States. It was a country, he now knew, where people could voice their views on political issues and politicians without any fear of persecution. The beauty of America, he said, was that this tolerance of each person's ideas existed without the coercion and climate of terror that had arisen in Europe. For Americans, Einstein thought, the freedom to express their own views was so precious that they would rather die than give it up.

Einstein's deep appreciation for America's core values of freedom of belief and expression motivated his cold public anger and dissent when, in the early 1950s, a

ABOVE: *Senator Joseph McCarthy in 1954 at the height of the "Red scare."*

rising "Red scare" led by Senator Joseph McCarthy and others led to unhinged loyalty investigations and what resembled witch hunts against those suspected of communist sympathies. Einstein—who was a democratic socialist but who abhorred the abuse of individual freedoms in communist systems such as Russia—tried to maintain a middle ground between those who were reflexively anti-American and those reflexively anti-Soviet.

After Einstein was given the Lord & Taylor award for the independent nature of his

> **"Look in the mirror and see how disgraceful you look without a haircut, like a wild man."**
> **— Sam Epkin of Cleveland**

"I AM AN AMERICAN DAY"

At the end of World War II, President Harry S. Truman had decreed a day which would honor all new American citizens. The judge who had taken Einstein's oath of allegiance as a new citizen in 1940, sent out thousands of invitations to attend a celebration at a park in Trenton to all those whom he had sworn in. Normally uninterested in such affairs, Einstein, remarkably, showed up at the party, bringing the rest of his household with him. He threw himself into the festive spirit, smiling amiably throughout the ceremony, while a young girl perched on his lap. His participation in "I Am an American Day" was perhaps a sign of the genuine pleasure that his status as an American citizen had given him.

BELOW: Einstein was not the only person to enjoy "I Am an American Day" as this picture of the O'Neill sisters in 1946 shows.

ABOVE, LEFT: *Social reformer and socialist politician Norman Thomas in 1947. Thomas and Einstein had first met in 1930 when the former tried to persuade Einstein that pacifism was not possible without economic reforms.*

ABOVE, RIGHT: *Drew Pearson at home in 1955. Pearson was strongly opposed to Senator McCarthy's actions and frequently spoke out against them.*

thinking, he gave a radio talk, which was heard by William Frauenglass, a Brooklyn schoolteacher. Frauenglass had been called before a Senate committee in Washington to give testimony about the perceived growth of communist influence in American high schools. He had refused to do so, and he wanted Einstein to support him.

In his reply, Einstein argued that the liberty to teach freely was now being threatened by reactionary politicians. In answer to Frauenglass's question as to what teachers and intellectuals should do, Einstein declared that noncooperation along the model of India's Mahatma Gandhi was the only valid form of revolutionary activity. He advised anyone asked to testify before the Senate committee to refuse. Einstein told Frauenglass he could make the letter public.

Many would not have dared adopt this position at a time when inquisitions into

EINSTEIN'S AWARD

In 1953, as the "Red scare" was in full force, Einstein won an award that was given each year by the Lord & Taylor department stores. It was given for independent thinking, not something that was much celebrated in the early 1950s. Einstein won it for his scientific "nonconformity"—a trait that had often helped him in his career, and he was proud of it. In the radio talk he gave after his acceptance of the award, Einstein noted with great satisfaction that his obstinacy and nonconformism had finally been officially rewarded. He also could not resist adding a political point. He criticized the suppression of independent thought that was being engendered by Senator McCarthy's investigations.

RIGHT: Along with ordinary people like William Frauenglass, several high-profile people were also asked to testify before the House Un-American Activities Committee. Although Arthur Miller (right) testified in 1956 about his own activities, he refused to name anyone else.

BELOW: Einstein and Oppenheimer at the Institute for Advanced Study in December 1947.

Ich danke Ihnen für Ihre Aufklärungen. Mit dem „rechte Seite" meinte ich die theoretischen Grundlagen der Physik. –

Das Problem, vor welches sich die Intelligenz dieses Landes gestellt sieht, ist ein sehr ernstes. Es ist den reaktionären Politikern dieses Landes gelungen, durch Vorspiegelung einer äusseren Gefahr das Publikum gegen alle intellektuellen Bemühungen misstrauisch zu machen. Auf der Basis dieses Erfolges sind sie daran, die freie Lehre zu unterdrücken und alle nicht Gefügsamen aus ihren Stellungen zu verdrängen, d. h. auszuhungern.

Was soll die Minderheit der Intellektuellen thun gegen das Übel? Ich sehe offen gestanden nur den revolutionären Weg der Non-cooperation im Sinne Ghandi's. Jeder Intellektuelle, der vor ein Haus der comité's geladen wird, müsste jede Aussage verweigern, d. h. bereit sein, sich einsperren und wirtschaftlich ruinieren zu lassen, kurz, seine persönlichen Interessen den kulturellen Interessen des Landes zu opfern.

Wenn nur genug Personen finden, die diesen harten Weg zu gehen bereit sind, wird ihnen Erfolg beschieden sein. Wenn nicht, dann verdienen die Intellektuellen dieses Landes nichts Besseres als die Sklaverei, die ihnen zugedacht ist.

A. Einstein

Diese Verweigerung dürfte aber nicht gestützt werden auf den bekannten Trick der möglichen Selbstinkriminierung sondern darauf, dass es eines unbescholtenen Bürgers unwürdig ist, sich solcher Inquisition zu unterziehen, und dass diese Art der Inquisition gegen den Geist der Verfassung verstosse.

P. S. Dieser Brief ist nicht als „vertraulich" zu behandeln.

the loyalty of academic staff and others in public life had become widespread. Many professors were agitated, intimidated, uncertain, or unnerved. But Einstein had a long history of—and derived some satisfaction from—standing in opposition to prevailing orthodoxies. So he was both serene and stubborn during the McCarthy era. His approach was simple: he advocated noncooperation. His memory of how most German intellectuals had stood aside as the Nazis seized power strengthened him in the conviction that now was not the time to sit idle while intolerance took a hold in the United States.

As well as receiving vitriolic treatment in the newspapers when his letter to Frauenglass became public, Einstein was also swamped with hate mail from people he didn't know. A George Stringfellow of East Orange, New Jersey, claimed falsely: "Don't forget that you left a communist country to come here where you could have freedom. Don't abuse that freedom sir." Sam Epkin of Cleveland wrote: "Look in the mirror and see how disgraceful you look without a haircut, like a wild man, and wear a Russian wool cap like a Bolshevik." And Victor Lasky, the noted columnist and bane of the American left, penned a handwritten note: "Your most recent blast against the institutions of this great nation finally convinces me that, despite your great scientific knowledge, you are an idiot, a menace to this country."

But Senator McCarthy's own response was muted. He seemed somewhat intimidated by Einstein's stature. "Anyone who advises Americans to keep secret information which they have about spies and saboteurs is himself an enemy of America," he said, avoiding attacking Einstein head on.

There were also letters in support of Einstein. The philosopher Bertrand Russell, an old friend and comrade from Einstein's pacifist days, wrote an amusing response to the New York Times. "You seem to think that one should always obey the law, however bad," he told the paper. " I am compelled to suppose that you condemn George Washington and hold that your country ought to return to allegiance to Her Gracious Majesty, Queen Elizabeth II. As a loyal Briton, I of course applaud this view; but I fear it may not win much support in your country." In his reply, Einstein showed his gratitude for Russell, but complained that the whole of the intellectual community was now thoroughly cowed.

William Frauenglass's young son, Richard, also wrote, and Einstein found the letter so sweet that he kept it in his desk. "In these troubled times, your statement is one that might alter the course of this nation," the boy said, a sentiment which contained an element of truth. He added that he would keep Einstein's letter for the rest of his life and then added a postscript: "My favorite subjects are your favorite too—math and physics. Now I am taking trigonometry."

Einstein also became involved in the case of J. Robert Oppenheimer, who by the early 1950s, had become the head of the institute in Princeton where Einstein was ensconced. Oppenheimer had previously been in charge of the scientific team which developed the atom bomb, and he still had his security clearance as an advisor to the Atomic Energy Commission. He was a vulnerable target. He had publicly opposed the construction of the hydrogen bomb, which President Truman had approved, and his wife and brother had been members of the Communist Party before the war.

The result was that some in the government launched a closed set of hearings in 1953 with the goal of stripping away Oppenheimer's security clearance. It was a symbolic fight rather than a substantive one, since the clearance was in any case soon due to expire. Yet in the polemical atmosphere of the times, neither side wanted to back down.

Einstein thought Oppenheimer was a "fool" for placing himself in this position. He had served the United States well, and ungrateful though the authorities might seem, he should not go along with the "witch hunt"

LEFT: *Einstein's reply to William Frauenglass's letter in which he recommended that Frauenglass follow Mahatma Gandhi's example of nonviolence and noncooperation. (See Translations page 157.)*

> "Anyone who advises Americans to keep secret information which they have about spies and saboteurs is himself an enemy of America."
> —Senator McCarthy

and would be better advised simply to resign. Einstein commented that Oppenheimer's seeming infatuation with the United States government was not reciprocated. His advice to Oppenheimer, he told a colleague, was that he should make his way to Washington without delay, tell the officials in no uncertain terms that they were buffoons, and then head home.

Although the hearings were supposed to be secret, they leaked and became a public issue. The Atomic Energy Commission voted Oppenheimer to be, in their view, a loyal American. But then it went on to declare him a security risk, and revoked his clearance just the day before it was due to expire. When a group of faculty members at the Princeton Institute began a petition in support of Oppenheimer, Einstein had no hesitation in signing it. When some members of the faculty demurred, fearing what would happen to them if they did put their signatures to it, Einstein was spurred into action. He "put his 'revolutionary talents' into action to garner support," recalled one of his friends. It did not take him long to cajole or embarrass the entire faculty into signing the petition.

To Einstein, the McCarthy-era loyalty trials and witch hunts were reminiscent of the rise of fascism in Germany. He considered that the greatest threat to America, as he

LEFT: *Edward R. Murrow on March 12, 1954 defending his attack on McCarthy that had taken place three days earlier on his program* See It Now. *This broadcast played a vital part in turning public opinion against McCarthy.*

explained to the Socialist leader Norman Thomas, would come from those who exploited the fear of communism to abolish precious civil liberties and not from the communists themselves.

There were a few people in Princeton who tried to keep Einstein from speaking out. They were afraid of the adverse effect his outspokenness might have on the institute or, indeed, on Einstein personally. He retorted that it was concerns such as this that made his hair turn gray. He took an almost boyish delight in saying just exactly what he chose. He wrote to his old friend the Belgian queen mother to say that he had become a kind of *cause célèbre* in his adopted country, principally owing to his inability to remain quiet in the face of such unacceptable developments. He believed that it was the duty of the older generation, who had correspondingly less to lose, to speak out for the young, on whom the pressures to conform were far greater.

Einstein's worry that McCarthyism was a precursor to an American plunge into the horrors of fascism was understandable, given what he had witnessed in Europe. Not used to the passions of a democracy, he did not realize that it would turn out to be just another of the fast-moving fashions in political sentiment. He did not understand that the United States' political system was strong enough to resist the battering and that its instinctive protection of individual liberty would endure. American democracy ended up surviving, as it always had. Senator McCarthy's reputation was undone in a series of hearings in a case against the army and by his own colleagues in the Senate, President Eishenhower, and crusading journalists such as Drew Pearson and Edward R. Murrow. At the end of the "Red scares," both Einstein and Oppenheimer were still harbored safely in their Princeton haven, free to work and express themselves as they liked. Einstein was saved from dying an embittered man, and by the end his wry humor and detachment had been restored.

NEWSPAPER REACTIONS TO EINSTEIN'S LETTER TO FRAUENGLASS

There was a public uproar in the newspapers when Einstein's letter to Frauenglass was published. The *New York Times* intoned: "To employ unnatural and illegal forces of civil disobedience, as Professor Einstein advises, is in this case to attack one evil with another." The *Washington Post* commented: "He has put himself in the extremist category by his irresponsible suggestion. He has proved once more that genius in science is no guarantee of sagacity in political affairs." The *Philadelphia Inquirer* added: "It is particularly regrettable when a scholar of his attainments, full of honors, should permit himself to be used as an instrument of propaganda by the enemies of the country that has given him such a secure refuge .... Dr. Einstein has come down from the stars to dabble in ideological politics, with lamentable results .... His statement is shocking not only in its aid to the Communist haters of America, but also in its display of bad taste and bad manners." The *Chicago Tribune*'s opinion was equally biting: "It is always astonishing to find that a man of great intellectual power in some directions is a simpleton or even a jackass in others."

Farewell

In a letter to his son Hans Albert at the end of 1954, Einstein expressed his pleasant surprise at how America had suddenly grown tired of the "Red scares." He wrote how the United States had come to seem more and more alien to him, but that in the end the Americans had returned everything to its proper state. The country was a factory in which everything, even madness, was produced on an industrial scale. Yet the fickleness of fashion there dictated that even the most alarming developments turned out to be but passing fads.

Einstein had turned 75, and Hans Albert had just turned 50. They had long since reconciled, and Einstein even relished the joy of being a grandfather.

Einstein was officially retired, but he was allowed to keep his office at the Institute. He carried on, wandering in there every morning at a reasonable hour, and would spend part of the day wrestling with equations that he hoped would make some progress, no matter how slight, toward a unified field theory that would explain away the purported uncertainties at the heart of quantum mechanics. He had begun to marvel at the concept of fields (meaning continuous structures) back when his father gave him a compass 70 years before, and it had guided his theories ever since.

But a unified field theory was a horizon that seemed to be constantly receding. He would come to work many days with a new strategy, sometimes with a pile of

ABOVE: When Louis de Broglie wrote his doctoral thesis on the theory of electron waves in 1924, it proved to be so complex that it was passed to Einstein for evaluation. Einstein endorsed the thesis and he and de Broglie remained friends.

TOP: Einstein with his son Hans Albert and grandson Bernhard in 1936. Two years later Hans Albert and his family moved to America where Einstein's relationship with his eldest son continued to flourish.

11.5.54.

Lieber Albert.

Die Ehrlichkeit verlangt es zu gestehen, dass Frieda mich an Deinen 50. Geburtstag erinnert hat. Und ich bin ihr dankbar dafür. Denn man hat nur bei solchen Gelegenheiten die Möglichkeit auszudrücken, wie man fühlt. Sonst scheut man sich.

Es ist mir eine Freude, einen tüchtigen Sohn zu haben, der die hauptsächliche Geste meines eigenen Wesens geerbt hat: sich erheben über das blosse Dasein, indem man seine besten Kräfte durch die Jahre hindurch einem unpersönlichen Ziel widmet. Dies ist ja das beste, ja das einzige Mittel, durch das wir uns vom das persönliche Schicksal und von den Menschen unabhängig machen können. Bei Dir ist es die Untersuchung der Vorgänge, die die Gestaltung der Wasserläufe bestimmen. Seit dem Verlassen der Schule hat es Dich nicht losgelassen, sodass Du nun auf eine kompakte Leistung zurückschauen kannst. Dies ist es, was einem Befriedigung gibt und dem Leben einen Sinn.

Gemeinsam ist uns auch das unablässige Grübeln und die Abneigung gegen das Bel-Studieren von Literatur. Das ist zwar ein Laster aber für unsereinen ein unvermeidliches. Es ist eine eigensinnige und gewissermassen heroische Art der intellektuellen Existenz.

Oft erinnere ich mich an besonders charakteristische Sachen, die Du Dir in der Kindheit geleistet hast. So entdeckte ich einmal, als ich mich rasieren wollte, dass Du meine Rasiermesser heimlich zum Holz-Schnitzen verwendet hattest. Das Ding war zu einer Art Säge geworden. Auch kommen mir die Blüten aus Deiner Kindersprache in den Sinn, z. B. das Wort "Voio Voio." Es sollte ursprünglich Vorhang bedeuten, bezeichnete dann aber alles, was gross und eindrucksvoll aussieht, aber wenig Substanz hat, z. B. Rauch aus dem Kamin oder ein leerer Redeschwall.

Und nun bist auch Du schon bejahrt und sogar Respektsperson! Bleibt nichts, als herzlich Glück zu wünschen.

Trachte weiter wie bisher, Bewahre den Humor, sei gut zu den Menschen, aber mach Dir nichts aus ihren Worten und Taten. Dein Papa.

ABOVE: *Einstein's letter to Hans Albert on his son's 50th birthday. (See Translations, page 157.)*

paper scraps on which he had scribbled new equations. He would then go over them with his assistant, a female Israeli physicist. Einstein would ponder them. She would point out problems, which Einstein would try to resolve. The problems eventually overwhelmed each new strategy, but he was never disheartened, even as time began to run out for him, he would simply remark that the day's labors had been instructive.

Einstein occasionally poked fun at his own stubbornness. To the physicist Louis de Broglie, he mockingly likened himself to an ostrich burying its head in the sand in a desperate attempt not to be confronted by the reality of the much-reviled quanta. Given that he had developed his gravitational theories by keeping faith in a basic underlying principle, Einstein believed that, equally, the best way to find a unified theory was by clinging to his faith in a cosmos that was rigidly predictable and deterministic rather than plagued by probabilities, uncertainty, and a dice-playing deity. If that made him an ostrich, then so be it, he noted in a wry aside to de Broglie.

Einstein was also indefatigable in his search for a long-term guarantee of peace in the atomic age, a labor in which he was encouraged by the philosopher Bertrand Russell. Both of them had been vigorous opponents of World War I and yet had supported the Allied cause in World War II. It was, Russell commented, therefore now their duty to prevent World War III. "I think that eminent men of science ought to do something dramatic to bring home to the governments the disasters that may occur," wrote Russell. Einstein's response was to propose some "public declaration" to be made by themselves and a select group of other eminent scientists and intellectuals.

On his 76th birthday—March 14, 1955—Einstein was inundated with presents and visits by close friends. Oppenheimer came by with some mail and publications, and others came

RIGHT: *Einstein on his 75 birthday in 1954.*

EINSTEIN'S UNDELIVERED SPEECH FOR ISRAELI INDEPENDENCE DAY

A week before Einstein died, the Israeli Ambassador Abba Eban arrived at the house to discuss a radio address Einstein had agreed to deliver on the seventh anniversary of the Jewish state. Einstein had once been dubious about the idea of a Jewish nation-state—given his aversion to nationalism—but now he told Eban that he regarded the foundation of the State of Israel as one of the few political developments he had seen that had a truly moral quality. Einstein told Eban that he wanted to broaden his speech to speak of the need for peace and for the creation of a world government in the atomic age. Einstein never completed the speech. The draft lay unfinished by his bed on the day of his death. He began by invoking the universal spirit of humanity, speaking not purely as an American, or as a Jew. He wanted everyone to know that he planned to speak not about Israel, but about the need for peace in the nuclear age.

ABOVE: *In the hours before his death, Einstein requested that the notes for his speech for Israeli Independence Day should be brought to the hospital so that he could continue working. When he died, this one page of notes was left at his bedside with the final sentence left unfinished. (See Translations, page 157.)*

to pay their respects. The fifth-grade students of the Farmingdale Elementary School in New York sent him a present of a tie, one imagines because they had seen pictures of Einstein, and believed he was in need of one. In his polite reply, Einstein said that it was so long since he had worn ties and cuffs that he had almost forgotten what they were. Reporters gathered in front of 112 Mercer Street and tried to get him to pose for a birthday picture, but he was not up for it. He stayed inside.

Michele Besso, his old friend and scientific confidant from his Zurich student days, died just the next day. Einstein wrote his family a condolence note, and in it he gave forth on the nature of time and death. He seemed, in some way, to know that he had very little time left to live himself. In his letter to the Bessos, he wrote that he considered that his friend's time had come just a little ahead of his own, and that in any case, science reveals that notions of what has passed away and

ABOVE: *Dr. Thomas Harvey, chief pathologist at Princeton Hospital, speaking to reporters about the autopsy he performed on Einstein. He was able to state that Einstein could not have been cured through surgery.*

BELOW: *A newspaper announcing the death of Einstein in 1955.*

ABOVE: *Hans Albert (right) leaving the hospital following his father's death with Dr. Otto Nathan, executor of Einstein's will. Less than 15 hours after death, Einstein was cremated in front of a small group of his family and friends.*

what is yet to come are all illusory.

As he entered the last week of his life, Einstein concentrated on a few issues he held most dear. He and Bertrand Russell had been working on a manifesto calling for world peace in the nuclear age, and Einstein signed it on April 11. "We have to learn to think in a new way," the manifesto declared. "We have to learn to ask ourselves, not what steps can be taken to give military victory to whatever group we prefer, for there no longer are such steps; the question we have to ask ourselves is: what steps can be taken to prevent a military contest of which the issue must be disastrous to all parties?" It was from this document that the Pugwash Conferences would develop, annual gatherings at which scientists and thinkers would exchange ideas on how to promote nuclear disarmament.

On April 12, when he was at work at the Institute, Einstein felt a pain in his groin. His assistant asked if everything was alright. He replied that things were fine, but that he was not. An aneurysm, or blister, on his aorta had started to burst, and a group of doctors who came to his home told him that his only hope, albeit a slim one, was to have immediate surgery to repair the aorta. Einstein would have none of it. He told Dukas he considered the extension of life by artificial means to be distasteful. He felt that he had done his due, and that, when the time came, he would go with grace.

The next day, as the pain worsened, he was taken to Princeton hospital. His stepdaughter Margot told Hans Albert of the crisis, and he caught a plane from San Francisco to be with his father. On Sunday April 17, Einstein woke up feeling well enough to do some more work. Dukas brought him his glasses, papers, and pencils, and he proceeded to work on yet another attempt to find a unified field theory, filling pages with lines of equations. At one point he pointed to what he had written and complained, half in jest, to Hans Albert, that what he really needed was more mathematics.

Then the pain became too great. His family left, and eventually he went to sleep. In the middle of the night—around 1 a.m. on Monday, April 18, 1955—he began stirring. His nurse heard him utter a few last words in German. His aneurysm, in effect a big blister, had finally burst, and he was dead.

On the table beside him were 12 pages of equations. To the very end, he had kept up the struggle that he had begun on the day his father gave him a compass, the struggle to visualize all of the forces of the universe and to paint them using the brushstrokes of mathematics and imagination. When he got to the final page, his handwriting had become a bit unsteady, and there are a few mistakes in the arithmetic that he had crossed out and corrected. Yet before

Farewell 153

PRESS RELEASE   July 9, 1955

    The accompanying statement, which has been signed by some of the most eminent scientific authorities in different parts of the world, deals with the perils of a nuclear war. It makes it clear that neither side can hope for victory in such a war, and that there is a very real danger of the extermination of the human race by dust and rain from radio-active clouds. It suggests that neither the public nor the Governments of the world are adequately aware of the danger. It points out that an agreed prohibition of nuclear weapons, while it might be useful in lessening tension, would not afford a solution, since such weapons would certainly be manufactured and used in a great war in spite of previous agreements to the contrary. The only hope for mankind is the avoidance of war. To call for a way of thinking which shall make such avoidance possible is the purpose of this statement.

    The first move came as a collaboration between Einstein and myself. Einstein's signature was given in the last week of his life. Since his death I have approached men of scientific competence both in the East and in the West, for political disagreements should not influence men of science in estimating what is probable, but some of those approached have not yet replied. I am bringing the warning pronounced by the signatories to the notice of all the powerful Governments of the world in the earnest hope that they may agree to allow their citizens to survive.

                                                           - BERTRAND RUSSELL

==

LETTER TO HEADS OF STATE

Dear. . . .

    I enclose a statement, signed by some of the most eminent scientific authorities on nuclear warfare, pointing out the danger of utter and irretrievable disaster which would be involved in such warfare, and the consequent necessity of finding some way other than war by which international disputes can be settled. It is my earnest hope that you will give public expression to your opinion as to the problem dealt with in this statement, which is the most serious that has ever confronted the human race.

                                          Yours faithfully,
                                          BERTRAND RUSSELL

ABOVE: *The press release for the Einstein–Russell Manifesto in which Einstein and Bertrand Russell appealed for nuclear disarmament.*

he went to sleep that final time, he had persevered to the end of that page, filling it with line after line of equations, still hoping that he might get just that little bit closer to the basic law that underlies the universe.

It had been a long journey for the rebellious third-class patent clerk to a place where he dared to seek to read God's mind and unlock the most secret doors of his creation. He marveled at things that the rest of us do not usually focus on, such as exactly what pushes the compass needle to point north. He challenged postulates that more educated scholars took as a given, such as the assumption that time marches along second by second no matter how we observe it. He visualized equations and the reality they painted, such as when he figured out that the mathematical contrivance that Max Planck had put into an equation meant that light came in particles as well as waves. He did thought experiments that were at once amazingly imaginative and simple: riding alongside a light beam or how a person on a moving train would see the timing of lightning strikes differently from a person on the platform. And he made connections that were so glaringly obvious that no one had seen them before, such as that the effects of acceleration were equivalent to the effects of gravity. Without Einstein, the world as we know it would not exist.

EINSTEIN'S LEGACY

Decades after his death, all of Einstein's great discoveries remain durable, and we are still living in his universe, one defined on the macro scale by his theory of relativity and on the micro scale by his quantum theory. His fingerprints are all over the technologies that have defined our times, from lasers and DVDs to atomic power and fiber optics, to space travel and even semiconductors. From the infinite to the infinitesimal — from the largest concept imaginable, the expansion of the universe, to the very smallest one, the emission of photons from the nucleus of an atom — Einstein's creativity continues to define the vast sweep of what we know about our cosmos and everything in it.

ABOVE: *Without Einstein, inventions such as this laser would not exist.*

Translations

Page 21 Graduation Certificate

The Board of Education at Kantons Aargau certifies that: Mr Albert Einstein from Ulm, born on 14 March 1879 attended third and fourth classes at the Aargau Kantons technical school. Having taken both the written and oral leaving examinations that took place on 18, 19, 21 and 30 September 1896 he received these marks:

German Oral and Literature	5
French	3
English	-
Italian	5
History	6
Geography	4
Algebra	6
Geometry (including Planar Geometry, Trigonometry, Spatial Geometry and Analytical Geometry)	6
Descriptive Geometry	6
Physics	6
Chemistry	5
Natural History	5
Artistic Drawing	4
Technical Drawing	4

Because of this, on this date the student is hereupon granted the school-leaving certificate.

Aarau 3 October 1896

Signed by the school president and the secretary

Page 29 Postcard

Mr. Albert Einstein,
18.II Tillierstrasse,
Berne, Switzerland

Leopoldskron Castle, near Salzburg
Bettenhausen, Salzburg

Dear (illegible),
I am already in Budapest. Things are moving fast, but it's tough, things are going really badly for me. What are you getting up to? Write to me soon, won't you? Your poor Dollie.

Pages 58–59 Letter to Elsa

Dear Elsa

Thank you very much for your letter. It was kind of you to have thought of me. There is no book on relativity that would be understandable to a lay person. But what do you have a relat. cousin for? If you are ever in Zurich, then we (without my wife, who is sadly very jealous) will go for a walk, and I will talk to you about those things I have discovered at that time. At the moment I am working on an extension of the theory, which is very difficult. I will try to find a picture of myself for you. I would prefer to come myself, but I am burdened with so much work that I am happy if I succeed in keeping up even though I work all day without taking any rest. Nobless oblige – this fame comes with a certain amount of unhappiness.

If you would really like to make me happy, then you should spend a couple of days here sometime.

With best regards, also to your children, your
Albert

Page 65 Haber Letter

To: Professor Einstein, Chair of theoretical physics at German University of Prague
From: Haber, Kaiser Wilhelm Institute for physical chemistry and electro-chemistry
Date: 19th December, 1911

My dear colleague,

Sorry not to have written earlier. It's not that I wasn't grateful for your letter, it's more that it fired my ambition. For you were so friendly in your letter and told me so much that I considered myself obliged to make an effort of my own to justify your faith in me. You will now be able to read the modest results of my labors in the next edition of the Proceedings of the German Physics Association. I summarize the main points below:

A coulomb force is introduced into the state equation of the solid body. The value for the electrostatic charge of the lone electron is correctly calculated from the compressibility and the atomic volume if we assume that this coulomb force is opposed to the compression. The solid body is an electron grid, in the meshes of which the charged positive particles hang. The linear oscillation of the electrons in this grid, separated into two half-amplitude circulars, supplies the diamagnetism when influenced by an external magnetic field. The order of magnitude can be obtained precisely, and the size with pretty good accuracy, if we derive the susceptibility using this idea and making the assumption that the maximum amplitude is comparable with the interval between the centres of two atoms. The essential point here, as in all these observations, is that the ratio of the maximum amplitude to the atom's diameter is universal. From the frequency of the selective photo-electric ion, and using this notion, we can also calculate, reasonably accurately, the value of the paramagnetic saturation. The root law can be combined with your compressibility law and with the Lindemann formulae in an integrated manner via the so-called image of the electrical solid body. The variable hv is identical to the electrostatic potential of the electron in the lattice of the solid substance multiplied by the electron charge, until an error arises due to the temperature. In all the examples for which I have carried out the calculations, and provided the numerical values have been correctly chosen, the variable hv also matches the heat tone, within 3%. We generally do not have the temperature function; if we could include it, I believe this would enable us to overcome the difficulties completely. From the theoretical point of view, I consider it a great deficiency that I did not know the energy equation for an oscillator, the frequency of which is dependent on the temperature. This is the characteristic of most natural oscillators, at least of the solid ones. It would be extremely useful to know this equation. Due to these same factors, I haven't yet finished deriving the thermal effect according to Richardson, but I am certain that I'll soon be able to demonstrate how this effect arises from quantum theory and from the notion of the electrical solid body.

I've had an enormous lot to learn, and I've absolutely exhausted my powers. But there's nobody I'm as grateful to as you for your teaching, and for this I send you my most sincere and most heartfelt thanks. And I close with the request that you will send me your comments and so instruct me further. I've taken the liberty of using you as a reference point many times – and in particular regarding the conversation which you were kind enough to have with me in the home of our mutual friend Marx in Karlsruhe, which I found extremely enlightening.

Wishing you a very happy Christmas, I remain

Your friend,
Haber

Page 67 Relativity Notes

Above, left page

End of lecture at end of exercise book (noted down in Zurich).
General Relativity, Summer Term 1919. Berlin
5.V
Deficiencies of classical mechanics
Equality of heavy & inert masses remains unexplained (Eötvös)
Preference for inertial system remains unexplained.
Equivalence principle.
12.V
More on equivalence principle
[Equations]
Atom at point of origin also has frequency of V_1
(If there's time) Therefore clock at P, considered from point of origin, goes faster than clock at point of origin.
Red shift of spectral lines.
Measurement of time using clocks of an identical nature everywhere is not possible.
[Diagram and Symbols] light-time; [Symbols] speed gained [Equation]
Generalized (Equation)

Above, right page

Curvature of light beams demonstrates that speed of light is dependent on location.
[Diagram and Equation]
Through approximate integration [Equation]
Speed of light dependent on gravitational potential.
Correlates with observation above in that light clock [Equation]
Deflection on star. [Diagram and Equations]
Transfer of result to any gravitational fields hypothetical. It will be demonstrated later that result
not correct.
Uniform red system. Running of clock. Confirmation of result above. Invalidity of Euclidean geometry for scales.

Below, left page

No transmission with Fizeau's experiment [Equation] only [Symbols] of zero for various functions of $(x - Vt)$

Matter conditions as per x diff [Equations]
9.XI not applicable due to revolution
16.XI. Lorentz transformation
23.XI. Lorentz transformation
30.XI Rigid bodies and clocks
7.XII Addition theorem of speed. Minkowski's interpretation of the Lorentz transformation
14. XII The principle of relativity and the Lorentz transformation.
Vectors & tensors as supports for theory.
Theory of tensors

Below, right page

but here the $g_{\mu\nu}$ values are real functions of the location. Best as per earlier
gravitation field. Above we encountered a special case (Equation)
where c was variable. There is clearly no invariant for any substitution.
Interrelationship of $g_{\mu\nu}$ with gravitation field is demonstrated when minimal lines are considered (Equation)

1) special theory of relativity (Equations)
2) Special case above (Equations) for low speed
3) General (Equation)

The g_{uv} value is determined simultaneously by metrics (scales & clocks) and by gravitation field.

Page 76 Divorce Letter

Dear Mileva,

The desire finally to bring a certain amount of order into my private life impels me to suggest a divorce to you for the second time. I am firmly resolved to do everything to make this step possible. In the event of a divorce, I would guarantee you significant financial advantages by going a very long way toward granting your requests.

1) 9000 M instead of 6000 M, with the stipulation that 2000 of this per year should be deposited to provide for the children.

2) In the event of a divorce, and should I be awarded the Nobel prize, it would, a priori, be transferred to you in full. The interest would remain freely available to you. The capital would be invested in Switzerland and put into a safe place for the children. My payments as listed in 1) would then cease, and be replaced by a per annum payment such that, when the interest is included, the total would amount to 8000 M. In that case,
you would have 8000 M to dispose of freely.

3) The widow's pension would
be secured for you in the event of
a divorce.

Naturally, I would not be prepared to make colossal sacrifices like this except in the event of divorce by consent. If you do not agree to a divorce, from that moment on only 6000 M a year will go to Switzerland—not one centime more. Please let me know whether you agree and whether you are prepared to bring an action for divorce against me. I'd do everything possible here to make sure that this didn't cause you any problems or make things awkward for you.

I receive regular news of you and the children from friends. I'm glad you are no longer suffering from fever and have had no further attacks. Albert's letters give me extraordinary pleasure; I can see from them how well the lad is developing as regards his mind and his character. I hope that, following his long stay in the mountain air, Tete won't be too vulnerable to any adverse effects caused by the polluted city air, and that he will soon be back home and feeling stronger..

My regards to you and your sister,

Albert

Kiss the children for me.

Please write soon with your answer.

Page 82 Lorentz Telegram

Eddington found stellar shift at solar limb, tentative value between nine-tenths of a second and twice that. Lorentz.

Page 144 Frauenglass Letter

Dear Mr. Frauenglass

Thank you for your communication. By "remote field" I referred to the theoretical foundations of physics.

The problem the intellectuals of this country are faced with is extremely serious. The reactionary politicians have succeeded in instilling in the public suspicion of all intellectual pursuits by speaking of a danger from without. Having been so far successful they are now trying to suppress the freedom of teaching and to deprive those who refuse to submit of their positions i.e. starving them.

What should the majority of intellectuals do against this evil? I can see only the revolutionary way of non-cooperation as Ghandi did. Every intellectual who is called before one of the committees ought to refuse to testify. He must be prepared to go to jail and suffer economic ruin, in short, to sacrifice his personal welfare in the interest of the cultural welfare of his country.

The refusal to testify must be based on the idea that it is shameful for an innocent citizen to submit to such an inquisition and that this type of inquisition contravenes the spirit of the Constitution.

If enough people are ready to take this step they will not be successful. If not, then the intellectuals of this country deserve nothing better
than the slavery which is intended for them.

Sincerely yours,

A. Einstein

P.S. This letter does not need to be considered as "confidential."

Page 149 Letter to Hans Albert

11.05.1954

Dear Albert,

Honesty compels me to admit that Friedi reminded me of your 50th birthday. And I'm grateful to her for it. For it's only on such occasions that we have the chance to express how we feel. The rest of the time, we shy away from it. It's a joy to me to have a son of my own, who has inherited the most important feature of my nature: to rise above mere existence by devoting the best that's in you to a selfless goal throughout the years. This is certainly the best, indeed the only way that we can make ourselves independent of personal fate and independent of other people. With you it's research into the events which determine the formation of watercourses. This has dominated your life ever since you left school, so that now you can look back on your solid achievement. This is what brings satisfaction and gives a meaning to life. Another thing we have in common is that we can't stop thinking—and we're always having to keep up with the literature, though we wish we didn't need to. That is certainly a vice, but for people like us it's one we can't avoid. It's an intellectual achievement that requires obstinacy, and even a degree of heroism. I often remember especially characteristic things you did when you were a child. I recall for instance once, when I wanted to have a shave, I discovered that you'd secretly used my cutthroat razor to do some woodcarving. It looked like the blade of a saw! And I also recall some of the mistakes you made in your childish way of talking—for example, the word "Voio-Voio". It was originally meant to be "Vorhang" ("curtain"), but then it came to describe anything that looked big and impressive but wasn't very substantial—e.g. smoke from the hearth or a meaningless torrent of words.

And now you're a middle-aged man, a person of importance, as they say! It only remains for me to wish you a happy birthday. Keep up the good work, keep your sense of humor, be kind to people, but don't let anything they may say or do worry you.

Your father

Page 151 Speech notes

I speak to you today not as an American, nor as a Jew, but as a human being who seeks with the greatest seriousness to look at things objectively. What I seek to accomplish is simply to serve with my feeble ability truth and justice at the risk of not pleasing anyone.

The issue is the conflict between Israel and Egypt. You may consider this to be a small and insignificant problem and may feel that there are more important things to worry about. But this is not true. In matters concerning truth and justice there can be no distinction between big and small problems; for the general principles that decide the behavior of men are indivisible. Whoever is careless with the truth in small matters cannot be trusted in important affairs.

This indivisibility applies to political as well as moral problems; for little problems cannot be properly appreciated unless they are understood in their interdependence with big problems. And the big problem in our time is the division of mankind into two antagonistic camps: the Communist World and the so-called Free World. Since the significance of the terms Free and Communist in this context is not clear to me, I prefer to speak of a power conflict between East and West, although, the world being round, it is not even clear what exactly is meant by the terms East and West.

Essentially, the conflict that is present today is no more than an age-old struggle for power, once again presented to mankind in semi-religious garb. The difference is that, this time, the development of atomic power has burdened the struggle with a ghostly feel; for both parties know and admit that, should the argument degenerate into actual war, mankind is doomed. Despite this knowledge, statesmen in responsible positions on both sides continue to employ the well-known tactic of attempting to frighten and discourage their opponent by assembling superior military capabilities. They do so even though this policy has the potential for war and destruction. No statesman in a position of responsibility has dared to adopt the only course that offers any chance of peace, the course of supernational security, since for a statesman to follow such a course would amount to political suicide. Political passions, once they have been fanned into flame, claim their victims …

Index

(page numbers in *italics* refer to photographs and captions; **bold** refers to quotations)

A
Aarau School 19, *20, 21, 29*
Anderson, Marian 129
"*Annus Mirabilis*", see "Miracle Year"; 1905 papers
anti-Semitism 11, 19, 43, 55, 87, 87, 88, 91, 94, 124
astronomy 37, 46, 47, 68, 79–85, *79, 80, 81, 85*
atomic bomb 116–21, 122–5, *122*
atoms 35–6, 41–2, *41*, 44, 66–7, 68, 96, 134–5, 155, 156, 157

B
Ben-Gurion, David 137, *137*
Berkeley, George 35
Bernstein, Aaron 15, *15*, 16
Besso, Michele 20, 45, 48, 51, 69, 152
Bethe, Hans A. *125*
blackbody radiation 42, *42*, 45, *45*, 86
Blumenfeld, Kurt 86, *86*, 87, *87*
Bohr, Niels 94, 95, 96, 98, *98*, 100–1, *101*, 132, 134, 135
Bond, Horace Mann 128, *128*
Bond, Julian 128
Born, Max 69, *94*, 100
Boyle, Robert 5
Brandeis, Louis 89, 90
Brillouin, Léon *133*
Broglie, Louis de *148*, 150
Burlington House 85, *85*

C
Casadesus, Gaby *111*
communism 123, 129, 140–7, 148, 157
compass, Einstein's fascination with 9, 10, 13, 148, 153, 155
Copernicus 46, 47, *47*, 48, 50

D
Dewey, John 107
Diaghilev, Sergei 84
Dirac, Paul 66, *66*
Dollie, see Marić, Mileva
Dukas, Helen (secretary) 113, *113*, 126, *127*, 137, 153
Dyson, Frank 85

E
Eban, Abba 137–8, *138, 139*, 150
Eddington, Arthur 79, *79*, 81, *81*, 83, *83*, 94, 95
Edison, Thomas 90
Ehrenfest, Paul 83, *101*
Einstein, Abraham (paternal grandfather) 10
Einstein, Albert:
 birth 8, 9–10, *9, 11, 12, 13*, 19
 death 129, 150–3, *151, 152*

divorce 60, 68, 69, 70, 74–7, 76, 92, 157
early life 9, 10, *10, 11, 17*, 56
education 14–17, 18–20, *20, 21*, 22–5, 23, 26–8, *29*, 90
equations 42, 51, *51*, 69, 122, 153
FBI dossier 118, 126, 140
FDR letter *120–1*
first scientific paper 30
graduation 28, *29*
honorary degree 128, *128*
legacy 155, *155*
Lord & Taylor award 140–1, 142
love of music 13, *13*, 24–5, 57, 73, 88, 108, *108*, 111, *111*
marriages 26–7, *26*, 37, 56, 74, 77
Nobel Prize 44, 75, *76*, 92–5, 92, 94, 96, 157
papers, see 1905 papers; *papers by name*
as patent clerk 6, 24, 33, *35*, 36, 41–2, 52, 53, 54, 55, 62
religion and Jewish identity 16–17, 19, 71, 79, 87, 91, 102–5, 136–9, 150, 155
renounces German citizenship 19
"sausage crown" 37
scientific reputation secured 135
sister's death 11
teaching 54, 63–4, 67, 68–9, 113, 142, 157
U.S. citizenship 110–15, *110*, 126, 127, *127, 130–1*, 141
Einstein, Bernhard (grandson) *148*
Einstein, Eduard ("Tete") (son) 55, 60, 63, 70, 74, 77, 108, 109, *109*
Einstein, Elsa (cousin) (second wife) 56–7, *56, 58–9*, 60, 63, 70, 75, 77, 99, 108, 110, 112, 113, 115, 156
 death *113*, 114, *114*, *114*
 marriage 74, 77
Einstein, Hans Albert (son) 37, *39*, 52, 55, 60, 63, 66, 68, 70–3, 74, 77, *109*, 148, *148, 149*, 153
Einstein, Hermann (father) 9, 10, 17
Einstein, Ilse (stepdaughter) 56
Einstein, Jakob (paternal uncle) 14, 17
Einstein, Lieserl 33
Einstein, Margot (stepdaughter) 56, 99, 112, *114*, 126, *127*, 153
Einstein, Maria ("Maja") (sister) 10, 11, *11*, 18
Einstein, Pauline (mother) 10, *13*, 27, *27*
Einstein–Russell Manifesto *154*
Eisenhower, Dwight D. 147
electromagnetic field theory 4–6, *6*, 7
electrons 42–3, *44*, 68, 96, 98, 100, *148*, 157
E=mc² 42, *51*, 69, 122
Epkin, Sam 140, 145
"EPR" (Einstein, Podolsky & Rosen) 133
ether 6, 18, 48, 50, 51, 89

F
Faraday, Michael 4

"The Field Equations of Gravity" (Einstein) 69
Flexner, Abraham 110, *110, 112*
fluxions 5
49 Kramgasse, Bern 33, *34, 35, 39*, 41
Franck, James 94
Frankfurter, Felix 89
Franklin, Benjamin 92, 104
Frauenglass, William 142, *143, 144*, 145, *145*, 147
free speech 126–9
Freud, Sigmund 84, 107, *107*

G
Galileo 4, 4, 6, 47, 48, *48, 49*
Gandhi, Mahatma 142, *144, 145*
Gedankenexperiment, see thought experiments
Gehrcke, Ernst 94
Gödel, Kurt 63
Golden, William 123
Goldstein, Herbert 104, *104*
Göring, Hermann 116, *116*
gravitational attraction 5, 10, 62, 63, 88
Grossmann, Marcel 24, 36, 63, 68
Grün, David 137
Guillaume, Charles-Edouard 93, *93*, 94
Gullstrand, Allvar 92, 94–5

H
Haber, Fritz 65, 70, 71, *71*, 73, 157
Habicht, Conrad 36, *37*, 41, 42, 44, 51
Halsman, Philippe 124, *124*
Harding, Warren G. 89, 90, *90, 91*
Harvey, Thomas 152
Hecht, Selig *125*
Heisenberg, Werner 49, 50, *51*, 96, 101, *101*, 132
Heitler, Walter *132*
Helmholtz, Hermann von 27
Hilbert, David 63, *63, 63*, 64, 66, 68–9, 70
Hitler, Adolf 106–9, *106*
Hogness, Thorfin R. *125*
Hooke, Robert 5
Hoover, J. Edgar 118, *118*, 140
Hume, David 35, *35*, 37, 39, *39*
Hylan, John F. 88

I
Israel 55, 86, 136–9, *136, 139*, 150, *151*, 157

J
Jefferson, Thomas 104

K
Kelvin, Lord 6, *6*
Kindred, J. J. 89
Klein, Oscar 94
Kleiner, Alfred 54–5
Koch, Caesar (uncle) 18

Konenkov, Margarita *114*
Kreisler, Fritz 111

L
Lasky, Victor 145
Laue, Max 52, 53, *53*
Lenard, Philipp 28, *28*, 42–3, 94
light:
  speed of 15, 36, 44, 48, 50–1, 62, 89, 156
  waves 4, 6, 9, 18, 36, *43*, *45*, 47, 48, 51
Linbergh, Charles 116, *116*, 118
$L=mV^2$ 51
Lorentz, Hendrik *82*, 83, *83*

M
McCarthy, John 140, *140*, 142, *142*, 145, *146*, *146*, 147
Mach, Ernst 37, 39, *39*
magnetic fields 4, 10, 43
Manhattan Project 98, 122
Manhattan Seder 136
Marić, Mileva ("Dollie") (first wife) 26–8, *26*, 28, *29*, 33, *39*, 57, 63, 73, 156
  divorce 60, 68, 70, 74–7, *76*, 92, 95, 157
  ill health 33, 56, 108
  loneliness 56
  marriage 37
  pregnancies 30–1, *30*, 33, *52*, 55
Marshall, George 123
Martin, David 35
Maxwell, James Clerk 4, 6, 7, *7*, 14, *14*, 22, 47–8, 51
Mercer Street 113–14, *113*, 115, *115*, 152
Michelson, Albert 48
militarism 16, *16*, 19, 60, 70, *70*, 71, 72–3, *107*, 153
Miller, Arthur *143*
"Miracle Year" (Einstein) 35, *35*, 41–5, *41*, 47–51, *51*
molecules 35–6, 42, 44
Molotov–Ribbentrop Pact 118
Morley, Edward 48
Morse, Philip M. *125*
Motte, Andrew 5
Murrow, Edward R. *146*, 147

N
"The Negro Question" (Einstein) 129
Nernst, Walther 57, 60, 70, 72, *72*
Newton, Isaac 4, 5, 6, 10, 39, 42, 50, 62, 63, 64, 69, 83, 92, 96
  theories of 5
  1905 papers 35, *35*, 36, 41, *41*, 42, 43, *45*, 49, *51*, 52, 54, 75, 94
Nobel, Alfred 92, *95*, 123
non-Euclidean geometry 63
nuclear disarmament 153, *154*

O
O'Connell, William Henry 104, *105*, *105*
"On the Quantum Theory of Radiation" (Einstein) 96
On the Special and General Theory of Relativity (Einstein) 69
Oppenheimer, J. Robert *114*, *143*, 145–6, *147*, 150
Oseen, Carl Wilhelm 94, *94*, 95
Osler, William 94
Ostwald, Wilhelm 92

P
pacifism 60, 70, 79, 81, 106–8, 118, 122
Palestine 86–8, 119, 136–7, *136*
particles 6, 41–4, *41*, *43*, 52, 96, 101, 132–4, 155, 157
Pearson, Drew *142*
Penrose, Boise 89, *89*
Pernet, Jean 23
Pestalozzi, Anna *19*
Pestalozzi, Johann Heinrich 19, *19*
photoelectric effect 42, 43, 44, 94, 95
photons 42, 43, *43*, 44, 96–7, 155 (see also quanta)
Planck, Max 43–4, 45, *45*, 52, 57, 60, 70, 96, 101, 155
Podolsky, Boris 133, *133*
Poincaré, Henri 37
Pugwash Conferences 153

Q
quanta 41, 42, 43, *45*, 54, 94, 95, 96, 97, 150 (see also photons)
quantum theory 41–4, 50, 96–101, *97*, 134

R
relativity, theories of 15, 18, 35, 39, 47–51, 53, 54, 62–9, *62*, 65, 66, *81*, *82*, 83, 84, 89, 90, 92, 93, *96*, 157
Rojansky, Vladimir *133*
Roosevelt, Franklin D. ("FDR") 116, 117–18, *118*, 119, *119*, *120*, 122
Rosen, Nathan 133
Rosenfeld, Léon *132*, 133
Russell, Bertrand 126, 145, 150, 153, *154*
Rutherford, Ernest 98

S
Sachs, Alexander 117–19, *118*
Schneider, Ilse 83
Schoenberg, Arnold 84
Schrödinger, Erwin 134, 135, *135*
Seidel, Toscha 111
Sinclair, Upton 107
Sitter, Willem de *83*
Solovine, Maurice 36–7, *37*, *38*
Solvay Conferences *97*, *98*, *99*

soup kitchens 75, *75*
space-time, curvature of 64, *64*, 68, 69
Spinoza, Baruch 37, 104
Stravinsky, Igor *83*, 84
Stringfellow, George 145
Sustermans, Justus *49*
Swing, Raymond Gram 123
Szilárd, Leó 116–17, *116*, 118, *125*

T
Talmud, Max 16, 17
Teller, Edward 117
Thermodynamics, Third Law of 72
Thomas, Norman 126, *142*, 146
Thomson, J. J. 85
thought experiments 9, 15, *15*, 18, 19, 36, 42, 47–9, 62, 99, 132–5, 155
Truman, Harry S. 141, 145

U
uncertainty principle 132–3
unified field theory 79, 95–7, 123, 135, 148, 150, 153
Urey, Harold C. *125*

V
Viereck, George Sylvester 102, *103*, *103*
visualization 14–16, 19, 36, 43, 64, 153, 155
Völkischer Block 87

W
Washington, George 145
wave theory 43, *43*, *45*
Weber, Heinrich 18, 22, *22*
Weissko, Victor F. *125*
Weizmann, Chaim 86–7, *86*, 88, *88*, 90, 91, 137
Wheatstone, Charles 45
"Why Socialism?" (Einstein) 126
Wigner, Eugene 116, *117*
Williams, John Sharp 89
Winteler, Anna ("Mama") *20*, 25
Winteler, Jost 20, 25
Winteler, Marie (girlfriend) 19, *20*, 25, 26, 27
Winteler, Paul (brother-in-law) 11, 18
Winteler, Rosa 20
Wittgenstein, Ludwig 84
world federalism 20, 107, 123, 125, 136, 150
World War I 60, *60*, *61*, 63, 70–3, *70*, *71*, 75, *75*, *78*, 79, *79*, 81, 86, 107, 135, 137, 150
World War II 119, 123, 136, 141, 150
World War III 125, 150

Z
Zileri, A. S. *49*
Zionism 86–7, *86*, 87, 90–1, 136–7
Zurich Polytechnic 6, 17, 18, 20, 22–5, *22–3*, 26, 27, 30, 35, 36, 44, 57, 63, 68, 94, 152
Zurich University 22, 67, 86, 135, 157

Credits

Acknowledgments
The publishers would like to thank the following people for their valuable assistance in the preparation of this book:

The Albert Einstein Archives, the Hebrew University of Jerusalem, Israel: Barbara Wolff, Chaya Becker.

Franklin D. Roosevelt Presidential Library: Robert D. Clark, Karen Anson.

Museum Boerhaave: Mara Scheelings.

William Ready Division of Archives & Research Collections, McMaster University, Mills Memorial Library: Jannie Balt, Kim Scott, Rick Stapleton.

Robert D. Farber University Archives & Special Collections Department, Brandeis University: Sarah Shoemaker.

Picture Credits
The publishers would like to thank the following sources for their kind permission to reproduce the pictures in this book.

Akg-Images: 25, 36, 61, 63, /Bildarchiv Pisarek: 77

© The Albert Einstein Archives, the Hebrew University of Jerusalem, Israel: 58–59, 67, 76, 144, 149, 151.

Corbis: 89, 90, 91, 103, 116b, /Bettmann: 56, 62, 75, 94, 106, 108, 110, 111, 112t, 112b, 113b, 115, 116t, 117b, 118t, 119l, 119r, 125, 127, 128, 135, 136t, 137, 140, 143t, 141, 146, 152t, 153/Christie's Images: 35, /Hulton-Deutsch Collection: 72–73b, 74, 97, /Lucas Jackson/Reuters: 30, /Ted Spiegel: 95, / Sergey Konenkov/Sygma: 114, 117t, /Underwood & Underwood: 37, 88t, 88b, 113t, 152b, /Baldwin H. Ward & Kathryn C. Ward: 7, /Oscar White: 142l

Courtesy of the Albert Einstein Archives, the Hebrew University of Jerusalem, Israel: 12, 21, 38, 130–131, 139.

ETH Zurich: 52

Emilio Segre Visual Archives, American Institute of Physics: 132, / Courtesy of the Besso Family: 20t, /Leon Brillouin Collection: 133

Franklin D. Roosevelt Presidential Library: 120–121.

Getty Images: 4, 11, 13t, 16, 17, 31, 39t, 44, 45t, 53, 71t, 84, 85, 86, 92, 99, 100, 109, 148t, /AFP: 64, 150, /Philippe Halsmam/ AFP: 124, /National Geographic: 46, / Popperfoto: 3, 138, 96, front and back endpapers /Time & Life Pictures: 50, 82t, 83r, 104, 107 118b, 122, 129, 143b, 142r

International Solvay Institutes: 98t

IStockphoto: 155

Mary Evans Picture Library: 28t

Museum Boerhaave: 82b

Photos 12: 93, /Ann Ronan Picture Library: 6, 51, 85

Picture Desk: 78

Private Collection: 105

Robert D. Farber University Archives & Special Collections Department, Brandeis University: 29, 65

Scala Archives: 87

Science Museum/SSPL: 14, 41, 45bl, 48, 49, 69, 96

Science Photo Library: 24, 98b, /American Institute of Physics: 34, 66, 83l, 101t, /Martyn F. Chillmaid: 42, /Luke Dodd: 46t, /Mehau Kulyk: 40, 43 / Lawrence Lawry: 45br, /Royal Astronomical Society: 80–81

Topfoto.co.uk: 9, 13b, 32, 79, /Ann Ronan Picture Library/HIP: 10, 20b, / The British Library/HIP: 5, /The Granger Collection: 18, 26, 27, 33, 39bl, 136b, /Roger-Viollet: 148b, /Ullstein Bild: 8, 15, 19, 39br, 22–23, 28b, 54, 71b, 72t, 87, 101b

William Ready Division of Archives & Research Collections, McMaster University, Mills Memorial Library: 154.

Every effort has been made to acknowledge correctly and contact the source and/or copyright holder of each picture and Carlton Books Limited apologises for any unintentional errors or omissions which will be corrected in future editions of this book.

Publishing Credits
Project Editor: Victoria Marshallsay
Editorial Consultant: Jonathan Wells
Design Manager: Katie Baxendale
Senior Designer: Darren Jordan
Picture Manager: Steve Behan
Production Controller: Emily Noto